WHAT'S SO MYSTERIOUS ABOUT METEORITES?

O. Richard Norton • Dorothy Sigler Norton

Front and back cover photos courtesy Svend Buhl, www.meteorite-recon.com
Front cover: Sikhote-Alin meteorite
Back cover: Henbury meteorites

Photos © 2012 by O. Richard Norton and Dorothy Sigler Norton
unless otherwise credited
Illustrations © 2012 by Dorothy Sigler Norton

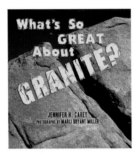 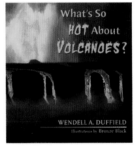

What's So Cool About Geology? is a registered
trademark of Mountain Press Publishing Company

Norton, O. Richard.
  What's so mysterious about meteorites? / O. Richard Norton, Dorothy Sigler Norton.
      p. cm.—(What's so cool about geology)
  Includes bibliographical references and index.
  ISBN 978-0-87842-591-4 (pbk. : alk. paper)
  1. Meteorites—Popular works. I. Norton, Dorothy S., 1945- II. Title.
  QB755.2.N67 2012
  551.3'97—dc23
                        2012014899

Mountain Press
PUBLISHING COMPANY
P.O. Box 2399 · Missoula, MT 59806 · 406-728-1900
800-234-5308 · info@mtnpress.com
www.mountain-press.com

O. Richard Norton passed away before he completed work on *What's So Mysterious About Meteorites*, a book that meant a lot to him. He had always wanted to provide a basic introduction to his beloved meteorites that could be read by young adults and anyone else interested in the subject of rocks that fall from the sky. While finishing it, I could hear his voice in my mind, explaining some detail about meteorites to students or to the many people who showed up at our door with boxes of rocks. He was a wonderful teacher, a great husband, and a generous friend to all who studied and searched for these curious rocks.

—DOROTHY SIGLER NORTON

It looks like a stained-glass window, but this colorful pattern is a thin slice of a meteorite, called a thin section, seen through a microscope. Light shining through the thinly sliced rock creates the colors that allow scientists to identify the minerals. This meteorite is from Mars. —Photo courtesy John Kashuba

ACKNOWLEDGMENTS

We would like to express our appreciation to the many people who provided assistance, materials, and encouragement for *What's So Mysterious About Meteorites?* At Mountain Press, publisher John Rimel and editor Jennifer Carey only nagged a little during the time it took us to complete the book. We'd also like to thank copy editor Jasmine Star and book designer Jeannie Painter. Ron Hartman, Dave Mouat, and Editor Emeritus of *Meteorite* magazine Joel Schiff graciously read early drafts and contributed thoughtful suggestions, as did Phillis and Tom Temple, Karen Shepard, Les Lambert, Don Bishop, and Fran Head. Rob Matson, noted asteroid and meteorite hunter, provided information about asteroids.

The photographs in this book come from many sources. Meteoriticists Steven B. Simon and Paul Sipiera furnished photographs and information about the Park Forest fall. We are grateful to the collectors who allowed us to photograph their meteorites—and "meteorwrongs." We especially want to mention pilot John Parker, who flew over Meteor Crater to take the panorama shot of the northern Arizona landmark, and John Kashuba, whose beautiful images of meteorites in thin section are as much art as they are science. Suzanne Morrison also took some wonderful pictures to our specifications. Howard Edin's photograph of a bright meteor is breathtaking. It is a pleasure to work with such fine photographers.

So, to all who made this book possible—scientists, photographers, meteorite hunters and dealers, astronauts and robot spacecraft exploring the cold expanse of space—and to all who continue to add to our knowledge of these mysterious ancient rocks, we say thank you!

Can you find the meteorite? Manuel, a resident of Berduc, Argentina, found one in a plowed field after a very bright fireball exploded and dropped stones on April 7, 2008. It's near his fingertips in the picture. —Photo courtesy Michael Farmer, www.meteoritehunter.com

CONTENTS

This spectacular meteor brightened the sky over Oklahoma on October 30, 2008. Do you see the constellation Orion in the lower center of sky? (Look for the three bright stars in a row that represent his belt.) —Photo courtesy Howard Edin

What's So Mysterious About Meteorites?

When you're standing outside on a clear day, gazing into the blue sky, it's hard to imagine anything up there except clouds and air. On a clear night, if the sky is very dark—away from big cities and night baseball games—you can see a lot farther. There are thousands of stars and that faint band of light called the Milky Way. Point binoculars or a small telescope at the Milky Way, and you'll see millions of stars, too many to count.

Every so often a bright light flashes across the sky. "Ooohh," you shout, "falling star!" It's an awesome sight. If it's really big and bright, it can leave a distinct trail behind it that lingers for a while like a ghost.

Of course it isn't really a star at all.

It's a tiny piece of rock or dust no larger than a grain of sand entering Earth's atmosphere and heating up. The rock or dust may have come from the tail of a comet that passed by long ago, leaving debris along its orbit. Comets are mixtures of rock and ice that probably formed in the outer solar system. As their orbits bring them close to the Sun they heat up. Their ice is vaporized and dust is released. Earth passes through clouds of cometary dust regularly. We see them at the same time every year: the Perseids in August and the Leonids in November, for example. Small particles burn up completely and never reach the ground. They are meteors. They occur in

1

meteor showers, and they're named for the constellations they always appear to come from, like Perseus and Leo.

You don't have to wait for annual meteor showers to see meteors. They can also appear sporadically, at any time, and from any direction. You can see them any night. They can be spectacular, but like shower meteors, most are very small, and they almost always burn up in the atmosphere.

But sometimes larger chunks of rock arrive from space. They create dazzling displays of light called fireballs, often with sound effects, and are seen and heard over a wide area.

Remember the silly little chicken who thought the sky was falling because an acorn hit her on the head? Chicken Little convinced her friends Henny Penny, Turkey Lurkey, and Cocky Lockey they were all in danger, and they hurried off to warn the king. On the way they were trapped by Foxy Loxy, who almost ate them all for dinner. The story ends well, with Chicken Little carrying an umbrella to protect herself from acorns after her big adventure. However, we shouldn't completely disregard Chicken Little's warning.

Sometimes rocks do drop right out of the clear blue sky.

About 50 tons of rocky material enters the atmosphere every day. Most of it is very small, like the small particles that produce meteors. If a larger piece makes it to the ground, it's called a meteorite.

Long ago, when people happened to see a meteorite falling from the sky, most of them probably thought it had been sent by the gods. Maybe it was frightening, especially if a lot of stones fell at once. Maybe it was seen as a warning or an omen. If the meteorite was made of iron, it was precious, because pure iron is rare and people could make it into tools and knives. Temples were built to honor meteorites in Greece and Turkey. Even as science replaced ancient superstitions in more modern times, people still found it difficult to accept the idea that a rock could just drop out of the sky. It took two widely witnessed events in the early 1800s—one in Connecticut and one in Normandy, France—to convince scientists that, indeed, meteorites are real. Still, very little was known about them because the equipment needed to study them in depth hadn't been invented. Some people thought they originated in storms or volcanoes on Earth. Looking into that beautiful blue sky, it's

Chicken Little was right.

hard to imagine how it could happen. Where did they come from? How did they get here? It was a mystery.

Today we know the answers to those questions, but there are still mysteries to be unraveled. Meteoritics, the science of meteorites, is quite young. New types of meteorites are still being found. They all contain important clues about the origin of the solar system, our Earth, and even life itself.

THE PARK FOREST FALL

It was a quiet, early-spring night on Indiana Street. In the south Chicago suburb of Park Forest, fourteen-year-old Robert Garza was asleep in his second-floor bedroom.

Shortly before midnight a deafening bang filled the air. Robert woke up to the sounds of shattering wood and breaking glass, followed by a loud crash and more breaking glass. Dazed and terrified, the boy curled up in his bed and called to his father. Noe Garza had heard the noise and thought someone was trying to break in. He hurried to check on his son.

—Photo courtesy Steve Witt

—Photo courtesy Steve Witt

—Photo courtesy Steve Witt

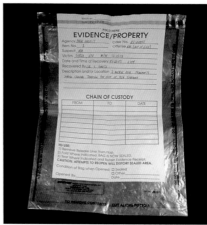

—Photo courtesy Greg Hupé

At first it wasn't clear what had happened. A large black and gray rock was resting on the bedroom floor. Incredibly, it had punctured the roof, breaking through plywood and a 2-inch-by-4-inch joist in the attic, then split into two pieces and punched through the ceiling. One piece dented the metal windowsill, destroyed the blinds, broke the window, and flew outside. The other piece bounced into a mirror on a closet door and smashed it. Cold air was blowing through the empty window frame, and pieces of ceiling plaster and broken drywall littered the floor and the bed. The only part of the room not damaged was the spot where Robert had been sleeping.

Noe Garza called the police. "Someone just threw a rock through my roof!" he told them. But who could throw a rock with that much force? The police took the rock away and put it in an evidence bag. Under "offense" they wrote, "N/A (Act of God)." The rock in the bag at the police station was soon joined by more mysterious rocks. One had hit the roof of the local firehouse. It quickly became clear that these trespassing stones were meteorites. As word got out, people started looking for them. And they found them all over the place.

Many people had seen the bright fireball as it streaked across the midwestern United States from southwest to northeast before the series of explosions that marked the end of its long journey. A few minutes later, low, booming sounds rolled across the sky like thunder. It had been raining lightly that night, but some people thought this thunder sounded all wrong.

When the fireball disappeared, rocks began to fall from the sky. Most of them landed in Park Forest and a few came down in nearby Olympia Fields and Matteson. They rained down on residential neighborhoods, damaging six houses, several cars, streets, and lawns. Hundreds of stones fell on that momentous night of March 26, 2003. It's very unusual for so many meteorites to fall in such a densely populated area. Remarkably, no one was injured. Fortunately, the Chicago area has several centers for meteorite studies and more than a few experts who could identify the rocks that had showered the area. By morning, television and newspaper reporters were on the scene and meteorite hunters and collectors were arriving from far and wide to search for celestial treasure.

Because so many scientists were nearby, they were able to collect samples and information quickly. Soon they would tell the world all about the rocks so many people

The largest stone from the Garza house weighs about 5 pounds. The cube measures 1 centimeter.
—Photo courtesy Greg Hupé

were looking for—and buying. Collectors and meteorite dealers were offering a lot of money for the stones. For weeks afterward, people scoured the area, sometimes on bicycles, sometimes crawling on the ground and peering under cars. Altogether, at least 66 pounds of meteorites were collected.

That might seem like a lot, but consider this: According to Steven Simon, Senior Scientist in the Department of Geophysical Sciences at the University of Chicago, the meteoroid that was the source of all these meteorites weighed at least 2,000 pounds—and possibly as much as 15,000 pounds—before it entered Earth's atmosphere. It's a good thing the pieces that survived the fiery journey through the atmosphere were small—from a few grams to 11.2 pounds. True, some buildings and cars were damaged, but imagine if the entire thing had made it to the ground intact. That could have been a real disaster.

After studying the Park Forest meteorites, scientists were able to say a lot about them. It turned out that they are examples of a fairly common type and had an interesting history in space.

Park Forest police chief Robert Maeyama and firefighter Dennis Kennedy retrieving an 18-ounce (515-gram) meteorite that was lodged in the roof of the fire station. The meteorite is now housed at the Field Museum in Chicago. —Photo courtesy Village of Park Forest and Steven Simon

The fire station meteorite sits next to the hole it made in the roof. —Photo courtesy © Paul P. Sipiera

A meteorite hit the center line on Park Avenue and produced this crater. —Photo courtesy Steve Witt

The yellow paint on this meteorite is evidence of its sudden impact with a fire hydrant. —Photo courtesy Steve Witt

The meteorites' arrival was well documented. Several police dashboard cameras caught the fireball on video. Orbiting satellites made optical and infrared observations. And the University of Western Ontario, 683 miles away, recorded audio and seismic measurements of the explosive events.

In space the meteoroid was about the size of a car. It hit the upper atmosphere at around 12 miles per second (over 43,000 miles per hour). At that speed, air pressure in front of the rocky body built up quickly and was much greater than the pressure behind it. As the air in front was compressed, it became very hot, and the meteoroid reached such a high temperature that its surface started to melt. It was so hot that part of the big rock was vaporized and became incandescent. The molecules of air around and behind it also became incandescent, producing the bright light and the glowing tail that stretched out for miles. Finally, the big rock broke apart in three separate explosive events at altitudes of 44 miles, 22 miles, and 16 miles above the ground.

Paul P. Sipiera and his daughter Paula (age nine) examining a hole that turned out to contain a nice Park Forest meteorite. She pointed it out to a meteorite collector. However, finders were not keepers because in the United States meteorites belong to the property owner. —Photo courtesy Steve Witt

Scientists at the University of Western Ontario who study ultra-low-frequency sound determined the fireball released energy equal to 500 tons of the explosive TNT. That's much less than an atomic bomb, but more than enough to destroy a car-size rock.

Scientists were even able to calculate the path the meteoroid was following when it was in space. It was typical of a special type of asteroid, called Apollo asteroids, which have orbits that cross the orbit of Earth. Most asteroids are very far away—beyond the planet Mars.

Does this mean meteorites are actually pieces of asteroids? What are asteroids anyway? And how can they get all the way from Mars to Chicago?

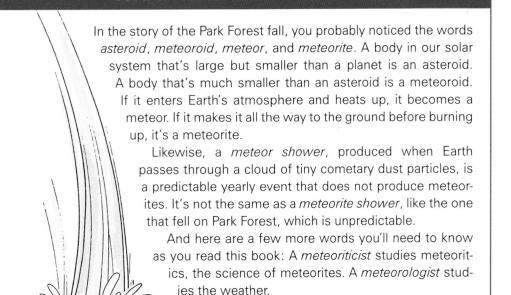

WHAT WAS THAT WORD AGAIN?

In the story of the Park Forest fall, you probably noticed the words *asteroid*, *meteoroid*, *meteor*, and *meteorite*. A body in our solar system that's large but smaller than a planet is an asteroid. A body that's much smaller than an asteroid is a meteoroid. If it enters Earth's atmosphere and heats up, it becomes a meteor. If it makes it all the way to the ground before burning up, it's a meteorite.

Likewise, a *meteor shower*, produced when Earth passes through a cloud of tiny cometary dust particles, is a predictable yearly event that does not produce meteorites. It's not the same as a *meteorite shower*, like the one that fell on Park Forest, which is unpredictable.

And here are a few more words you'll need to know as you read this book: A *meteoriticist* studies meteoritics, the science of meteorites. A *meteorologist* studies the weather.

Many scientific terms are based on Greek words. The words mentioned here come from *meteoros*, meaning "high in the air," *aster*, meaning "star," and *eidos*, meaning "shape" or "form."

Asteroid 951 Gaspra, imaged by the Galileo *spacecraft on October 29, 1991, from a distance of 3,300 miles. Gaspra is about 12 miles across. Note how its surface is pockmarked with craters of different sizes, the result of collisions with other asteroids.* —Photo courtesy NASA

Where Do Meteorites Come From?

Take a quick look at our solar system and you'll see something strange between the rocky inner planets—Mercury, Venus, Earth, and Mars—and the outer planets, the gas giants Jupiter, Saturn, Uranus, and Neptune. (Yes, sorry, Pluto is missing. It was demoted to dwarf planet in 2006.) At a distance of about 280 million miles from the Sun you find, instead of a planet, a zone of rocky and metallic chunks called the asteroid belt, and most meteorites come from that region. Some asteroids are huge, as big as mountains or even larger. Most are smaller. Many look like giant potatoes, lumpy and full of craters, but others have really strange shapes. No one knows exactly how many there are, but there are probably millions that are larger than a half-mile across and many more that are smaller. Astronomers using telescopes on Earth and space-based telescopes have discovered at least 540,000 of them. More than 260,000 have been numbered, and more than 16,200 have names. The largest resident of the belt is Ceres, named for the Roman goddess of agriculture. Before Ceres was discovered in 1801, thanks to a lucky accident by Giuseppe Piazzi, an astronomer in Sicily, these airless little worlds were completely unknown.

Not all meteorites come from asteroids, but we still have asteroids to thank for them. Some are rocks from the Moon or Mars that were sent flying into space during a collision with an asteroid. How can asteroids collide with planets and moons? Like the planets, asteroids in the belt follow regular orbital paths, and most of the time

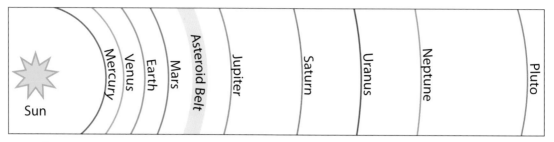

The asteroid belt, home to millions of asteroids, lies between the orbits of Mars and Jupiter.

everything stays where it belongs. The force that keeps planets and asteroids in their orbits is gravity, which causes objects to attract each other in proportion to their mass. The Sun, which is very massive, exerts a gravitational attraction on the planets even though they are very far away from it. Earth exerts a gravitational attraction on the Moon in the same way. On a smaller scale, we all know what happens when we drop a glass or wipe out on a bike. We experience gravity.

Things can change over time, though, because everything that has mass exerts some gravitational attraction on other things. The orbital paths of asteroids can change enough to send them crashing into each other and even into planets.

JUPITER AND THE ASTEROID BELT

When the asteroid belt was first discovered, people thought it must be the remnant of a destroyed planet. But what could destroy a planet? It's fun to imagine the answer to that question. We've all seen the movies and played the video games. Today we know there never was a planet in that location because the next one, Jupiter, exerts an enormous gravitational influence on the neighborhood. Astronomers think that when the solar system was first developing, Jupiter formed very quickly, in about 10 to 20 million years. Other small bodies were forming too, but they hadn't grown into a planet in the region where the asteroid belt is today. Jupiter's gravitational pull kept that from happening. Originally, there was probably enough material to make two Earth-sized planets. Now, thanks to Jupiter, there isn't enough in the whole asteroid belt to make a planet as massive as our Moon, which has only one-eightieth the mass of Earth. Most of the asteroid belt is simply empty space.

What makes Jupiter so powerful? It's huge. It's 1,382 times larger than Earth and weighs 317 times as much. Most asteroids in the belt have stable, nearly circular orbits around the sun. Some asteroids have been captured as moons, and some have undoubtedly collided with the giant planet. Others have been redirected into very elliptical orbits by Jupiter's powerful gravity. These new orbits could eventually send them heading for the inner solar system, toward the Sun or even Earth.

What else, besides the gravitational influence of Jupiter, can move an asteroid off its orbital path? Collisions, and lots of them. Even though the asteroid belt is not at all crowded, and these big wandering mountains of rock and metal aren't constantly

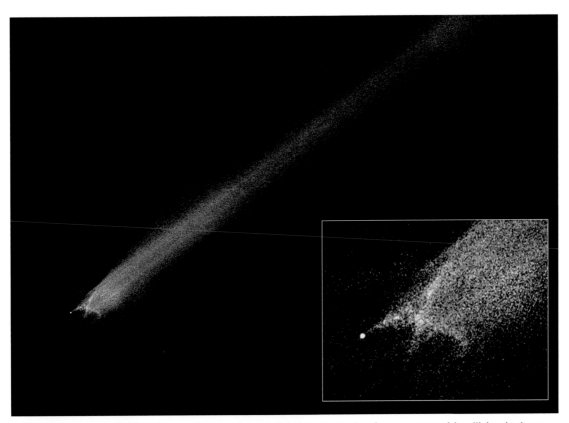

The Hubble Space Telescope made this picture of debris streaming from an asteroid collision in January 2010. Although we know asteroids have often crashed into each other with disastrous results, this is the first time it was witnessed and photographed. Scientists estimated the impact speed at more than 11,000 miles per hour. —Photo courtesy NASA, ESA, and David Jewitt (University of California, Los Angeles)

13

LOOKING AT ASTEROIDS WITH RADAR

You see it every day. The weather map on TV, the police officer writing a speeding ticket, the airplane landing safely at a busy airport—these things are all made possible by radar. Radar is one way NASA can observe asteroids and construct images of them. How does it work?

Radar detects an object at a distance by sending out, or transmitting, radio waves. Radio waves are invisible, travel at the speed of light, and are easy to detect, so they are perfect for this job. When radio waves bounce off an object, they are reflected back and detected by a receiver. The signal bouncing back allows measurements of speed, location, and sometimes shape.

Radio waves are similar to sound waves, which you're probably familiar with. You know what an echo is. Shout across a canyon or down a well and you'll hear your own voice coming back. How long it takes depends on how far away the canyon wall or bottom of the well is. You can also tell if something—like a freight train or an ambulance with its siren blaring—is coming at you or going away from you by the way it sounds. The sound is different coming and going because of something called the Doppler shift, after Christian Doppler, the physicist who proposed the reason for it in 1842. Sound travels in waves. The frequency of the sound waves from something that's moving is higher when the object is coming toward you and lower when it's moving away. The sound waves bunch up when the train or ambulance comes nearer because they are being emitted at shorter and shorter distances from you, the receiver. As the source recedes, they are emitted farther and farther away from you, and it takes longer for the sound to reach you. Radar uses echo and Doppler shift technology, but with radio waves instead of sound waves. The word *radar* stands for radio detection and ranging.

The Goldstone Apple Valley Radio Telescope in the Mojave Desert, California, is used by NASA to track asteroids.
—Photo courtesy NASA

bumping into each other, it can happen. Like our own Moon, most asteroids are covered with craters. They have a history of violent impacts, some quite recent. Even if there isn't a collision, the gravitational pull from another asteroid or a passing comet could be enough to send an asteroid in a completely new direction.

Collisions within the asteroid belt have formed distinct families of asteroids that travel together. At first broken pieces of both parent bodies form a tight cluster, but as the family ages the individuals spread out in their orbits and become separated—kind of like human families, except these asteroid families don't get back together for reunions or on holidays.

NEAR-EARTH ASTEROIDS

Asteroids that regularly travel through or come close to Earth's neighborhood are especially important to us because they're the ones that could potentially collide with Earth. They're called near-Earth asteroids, or NEAs. NEAs are found in three different groups, the Apollo, Amor, and Aten, each in their own orbit. The orbits of two of the groups—the Apollo and Aten asteroids—cross the orbit of Earth, and those are known as Earth-crossing asteroids. Earth-crossing may sound scary, but it just means their orbits overlap Earth's orbit, not that they are on a collision course. However, because they are constantly being influenced by the gravitational pull of Earth and other planets, our Moon, and each other, their paths can change. Even the energy from sunlight can affect an asteroid's motion over time.

The Apollo asteroids (named for the asteroid 1862 Apollo) and the Amor asteroids (named for the asteroid 1221 Amor) still spend part of their time in the asteroid belt. The Park Forest meteorites had a typical Apollo orbit. There is no way to determine when the meteoroid on its way to Chicago left its stable orbit in the asteroid belt and joined the wandering Apollo Earth-crossers. However, the texture of the meteorites shows a history of impacts and serious breakage and melting long ago. They came from a chunk of a broken world. They are like the pieces of a jigsaw puzzle, some light, some dark, with cracks running through them.

The Earth-crossing Aten asteroids (named for the asteroid 2062 Aten), have left their former home completely. They circle the Sun inside the orbit of Mars.

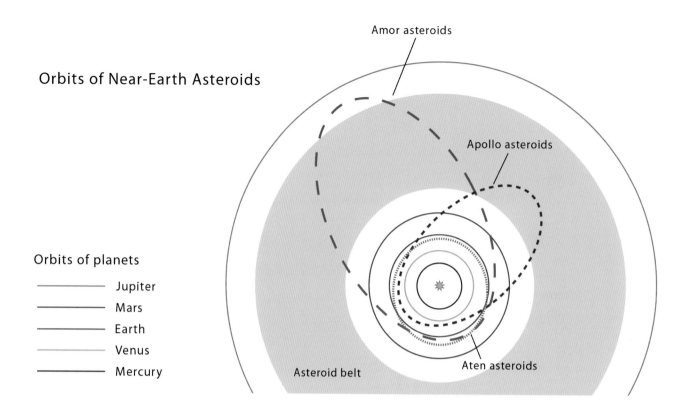

Orbits of Near-Earth Asteroids

Amor asteroids

Apollo asteroids

Orbits of planets

———————— Jupiter
———————— Mars
———————— Earth
———————— Venus
———————— Mercury

Asteroid belt

Aten asteroids

All the meteorites on Earth probably come from only about 140 asteroids—a pretty small sample of the millions that are out there. This means that pieces of a relatively few large asteroids in the asteroid belt have been blasted into space and found their way to Earth as near-Earth asteroids.

BEYOND THE BELT

Not all asteroids are found inside the asteroid belt. A few have been captured by planets and now circle them as moons. Phobos and Deimos, which are the moons of Mars, and some of the many moons of Jupiter, Saturn, and Uranus are probably captured asteroids. Two groups of asteroids, known as Trojans, travel with Jupiter in its orbit. Others have been found with the planet Neptune. Beyond Neptune lies a mysterious realm known as the Kuiper belt, named for astronomer Gerard Kuiper. There, frozen worlds exist that are similar to asteroids but made of ice. There may be another whole belt of orbiting objects out there.

AN ASTEROID BY ANY OTHER NAME

Ceres, the only dwarf planet in the asteroid belt, was thought to be a planet when it was discovered in 1801. Over the next six years, three more small bodies were found. Thirty-eight more years went by before the next one was located. Eventually, when astronomers realized they had discovered a new class of solar system object, they named them asteroids and devised a system of numbering them in the order they were found—1 Ceres, 2 Pallas, 3 Juno, 4 Vesta, 5 Astraea, and so on. By 1852, fifteen were known. Later, a new invention changed everything: photography. Though photography with cameras began in the early 1800s, it wasn't practical for astronomy until the 1890s. Once astronomers could take pictures of the sky, they made all kinds of new discoveries. For example, photos taken on consecutive nights could be compared to look for celestial bodies that had moved. By the end of the nineteenth century, more than three hundred asteroids were known.

Today, with computerized equipment on the ground and telescopes in space searching for asteroids, the number of known asteroids is huge. A new method of numbering and naming had to be devised. After World War II, the International Astronomical Union established the Minor Planet Center at the Smithsonian Astrophysical Observatory in Cambridge, Massachusetts. All data on new discoveries of asteroids and comets is processed there. A new find is first given a provisional number, which is based on the year and month of discovery. When the discovery is verified by further observation, a new number is assigned and the asteroid can be named by the finder.

You may be wondering how things are named in the solar system. The planets are all named for gods or goddesses in Greek or Roman mythology, except Earth, because originally no one knew Earth was a planet. Planetary moons are named for characters associated with those gods, like Phobos and Deimos, the sons of Mars. Craters on Mercury are named for poets and artists because Mercury was the god of communication. On Venus craters are named for women. The first asteroids were named for female figures in Greek and Roman mythology, such as Ceres, goddess of agriculture, but today the finder can name them for (almost) anyone or anything. John Lennon has an asteroid named after him. So do Charles Darwin, Ludwig van Beethoven, Stephen Hawking, James Bond, and British comedy group Monty Python. There is even a Mister Rogers.

Phobos, one of the two moons of Mars, may be a captured asteroid. Mars is named for the Roman god of war, and its moons are named for his sons (in Greek): Phobos, which means "fear," and Deimos, which means "terror." Phobos is 14 miles across.
—Mars Reconnaissance Orbiter image courtesy NASA/JPL-Caltech/University of Arizona

Officially, all asteroids qualify as minor planets if they haven't been captured by another planet to become a moon. The largest asteroids, which can be as big as Pluto, are also called dwarf planets, a name adopted by the International Astronomical Union in 2006. To be a dwarf planet, a body must orbit the Sun, be roughly spherical, and clear all other bodies from its orbit. The rest of the asteroids fall into the category of small solar system bodies.

ASTEROIDS OF MANY COLORS

You might think all asteroids are pretty much the same, but they aren't. Some are metallic, some are rocky, and some are a mixture of both. Some are light-colored, and some are very dark. Some have suffered through many violent events, and some have been completely shattered. Others haven't changed at all since the origin of the solar system. If you had a device called a spectrophotometer attached to a telescope, you could measure the light reflecting off the surface of an asteroid to help identify what it is made of. The spectra, or colors, of light reflecting off any surface, known

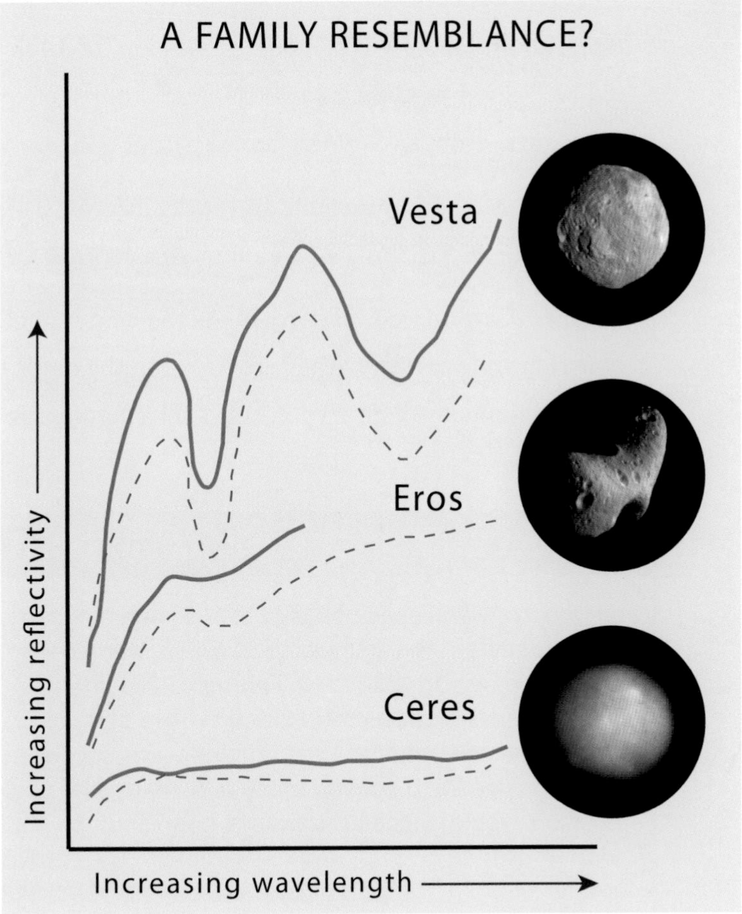

A FAMILY RESEMBLANCE?

Vesta

Eros

Ceres

Increasing reflectivity →

Increasing wavelength →

The spectra of three asteroids (red) compared with spectra of known meteorites (blue) are similar enough to suggest a relationship. —Images courtesy NASA

as the reflectance spectra, vary depending on what the object is made of. So far, the reflectance spectra of more than five hundred asteroids have been recorded and can be compared with light reflected off known meteorites. It's now possible to connect some meteorites with specific asteroid parent bodies, or at least say they are very similar. To know more, we will have to visit the asteroids in person or send out spacecraft to take samples.

Three basic types of asteroids are known based on their spectra and brightness as seen from Earth and the few spacecraft that have visited them:

- S-type—Silicaceous asteroids. Silicate minerals (a mixture of the elements silicon and oxygen) are the most abundant rock-forming minerals on Earth. Glass is pure silica. Typical beach sand is mostly pure silica. About 17 percent of the asteroids are rocky, S-type bodies. They are found in the inner part of the aster-oid belt, closer to Mars. Like most other stony meteorites, the Park Forest mete-orites came from an S-type asteroid parent body.

- M-type—Metallic or metallic and rocky asteroids. The metal is nearly pure iron with a little nickel. Only 8 percent of all asteroids are M-type.

- C-type—Carbonaceous asteroids. As the name suggests, these asteroids are car-bon-rich and very dark. They are found in the outer regions of the asteroid belt, closer to Jupiter. About 75 percent of all asteroids are C-type.

TWO LITTLE SPACECRAFT THAT COULD

Although several space probes have visited asteroids, only two have actually landed. The first was the *NEAR Shoemaker* spacecraft, which touched down on 433 Eros in February 2001. Then, in 2005, the Japa-nese *Hayabusa* spacecraft landed on the surface of 25143 Itokawa and collected samples. The samples were successfully returned to Earth in June 2010 after so many difficulties that no one thought the mission would succeed. It was a huge triumph for the engineers, who refused to give up. And the samples? They proved Itokawa is an S-type asteroid, the kind most meteorites come from.

NEAR Shoemaker *approaches Eros.* —Artwork courtesy NASA

ASTEROID SHAPES: TYPICAL AND SURPRISING

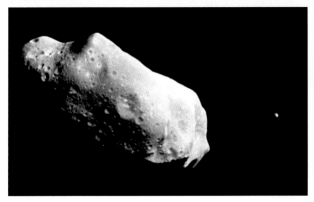

243 Ida, the first asteroid discovered with a tiny moon of its own. The moon is Dactyl. Ida is 35 miles long. After the discovery of Ida and Dactyl, more than 150 other asteroids with moons, or twin asteroid systems, were found. —Photo courtesy NASA

253 Mathilde, a 36-mile-wide C-type asteroid with a very dark surface, imaged by the NEAR Shoemaker *spacecraft in June 1999.* —Photo courtesy NASA

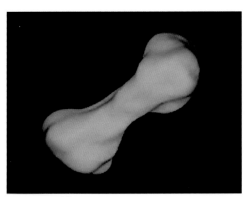

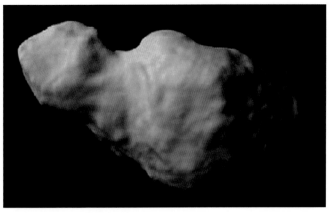

Radar image of 216 Kleopatra, a dog bone the size of New Jersey, apparently made of metal. —Image courtesy NASA/JPL/Northwestern University

Radar image of 4179 Toutatis, an Apollo asteroid only 2.9 miles long that frequently comes close to Earth. —Image courtesy NASA/JPL

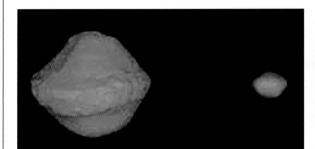

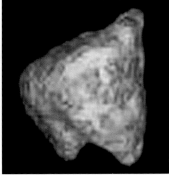

Radar image of 1999 KW4, a small double Aten asteroid system. The two bodies are actually orbiting each other. The big one (1 mile across) seems ready to fly apart at the equator. —Image courtesy NASA

Radar image of Apollo asteroid 6489 Golevka, only about 0.3 mile across. Doesn't this look like a giant tooth? —Image courtesy NASA/

This diamond in the sky is really 2867 Steins. It's about 3 miles across. —Image courtesy ESA ©2008 MPS for OSIRIS Team MPS/UPD/LAM/IAA/ RSSD/INTA/UPM/DASP/IDA

*A region in the Eagle
Nebula (M16), in the
constellation Serpens.
Deep within the
columns of interstel-
lar gas and dust are
knots and globules
of denser gas where
stars are being born.*
—Photo courtesy NASA/ESA

What Are Meteorites Made Of?

Like the asteroids most of them come from, meteorites are not all alike. Some are metallic, some are stony, and some are a mixture of both. To see why they are so different from each other, let's take a quick trip back in time.

The solar system formed 4.6 billion years ago from a nebula—a cloud of dust and gas in space. No one knows exactly how it formed, but it may have started with the violent explosion of a nearby supernova, a massive star at the end of its life cycle. The supernova released elements (substances made of only one kind of atom) into the cloud, and the expanding shock wave from the explosion disturbed and compressed the cloud, producing areas of turbulence. Material from the cloud collected in these areas, and they slowly became more massive. Eventually one got so dense that it became hot and started to produce light at its core. It became our Sun.

As dust and gas continued to gravitate toward the dense core and the young Sun, the nebula began to rotate and spread out in the shape of a disk. In the massive early Sun the temperature reached 3,600 degrees Fahrenheit. Yet at the outer edge of the disk it was a frigid minus 446 degrees Fahrenheit. These differences in temperature resulted in different chemical conditions across the disk. In the center the dust was vaporized. In cooler regions, the dust survived.

As time passed, turbulence caused clumps in the disk to collide and stick together. Like rolling snowballs, they collected more material. As they grew, they continued

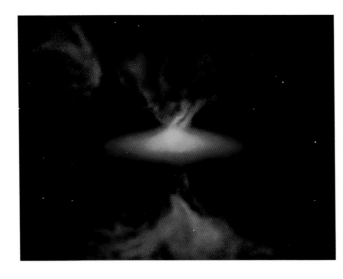

The rotating solar nebula flattens into a disk as its center gains mass and collapses to become the early Sun. —Painting by Dorothy Sigler Norton

SUPERNOVAS ARE NOT THE BIG BANG

Please don't confuse supernovas or the origin of our solar system, about 4.6 billion years ago, with the Big Bang—the beginning of everything—about 13.7 billion years ago. Our Sun and planets are relative newcomers in the universe. In fact, the Sun is a third-generation star. We know this because it contains heavy elements that it cannot produce itself—elements that can only be made when more massive stars (ten times the size of the Sun) explode as violent supernovas.

The cores of stars are giant thermonuclear furnaces, fusing the nuclei (centers) of hydrogen atoms to form helium. As hydrogen runs low, large stars begin fusing helium nuclei (at 100 million degrees Fahrenheit!) to form heavier elements, such as carbon. When the gravitational forces and heat from fusion can't keep temperatures high enough to continue, the star begins to collapse.

Large stars reach the point where they finally produce iron. This is the end stage for the star. When the biggest stars run out of fuel, they suddenly collapse and rebound in gigantic explosions that rip them apart—supernovas.

We have iron in our red blood cells and steel in our bridges because more than one earlier and very massive star lived and died, scattering heavy elemental particles into space.

to bump into each other and stick together to form bigger masses, a process called accretion. They grew larger and larger, and eventually the increasing gravitation of the larger masses swept up all the material in their neighborhoods. They became the planets we know today. Those closer to the Sun—Mercury, Venus, Earth, and Mars—accreted more iron and rocky material because the iron, calcium, and magnesium silicate minerals that make up the rocky planets condense out of gas at high temperatures. Farther out, where it was colder, the growing planets accreted some metal and rock but also attracted ice and the gases hydrogen and helium, which condense at lower temperatures. Reactions of hydrogen with carbon formed the gas methane, and reactions of hydrogen with nitrogen formed the gas ammonia. These gases make up the atmospheres of the giant gas planets Jupiter and Saturn. They have rocky cores but no surface you could walk on. Their moons, however, are made of rock and ice.

Some asteroids went through the same processes the early planets did. They accreted a lot of material. When they were big enough, gravity took over, as it always does. Their own increasing mass smoothed them from odd shapes into spherical or almost spherical forms. If the sphere was big enough, the whole thing melted. Impacts on the surface caused some melting, but most of the heat probably came from certain

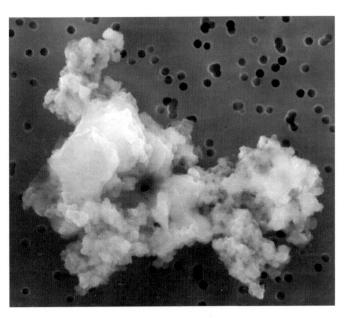

An interplanetary dust particle collected in Earth's atmosphere, probably from an asteroid or comet. Particles like this contain microscopic fragments from the original solar nebula from which our solar system formed. This one is 10 microns across, one-tenth the width of a typical human hair.
—Photo courtesy NASA

elements in the form of radioactive, or unstable, isotopes. These forms of elements change over time, decaying fairly quickly in the geologic sense (less than a million years) and releasing heat.

Eventually, if there is enough melting, iron and other metals in the mix start to sink toward the middle of the sphere and form a mostly iron core. The metals gravitate (literally) to the center because they are heavier than the surrounding minerals. The layer above the core is the mantle. The lightest minerals in the mix tend to float to the top and form a crust. The process that separates the heavy from the light is called differentiation.

Does this layered arrangement of core, mantle, and crust remind you of anything? If you said "Earth," you're right. Earth is differentiated too, and so are all of the rocky inner planets, as well as our Moon. Today, Ceres, Pallas, and Vesta are the only nearly spherical bodies in the asteroid belt large enough to be differentiated, but there must have been others. That's pretty apparent because there are asteroids made of metal and meteorites made of iron. They couldn't have formed without differentiation.

The bottom line is that meteorites have a lot in common with Earth's rocks. All the elements in meteorites are also found on Earth. Rocks are made of minerals, and the most common minerals in meteorites—olivine, pyroxene, iron-nickel—are also

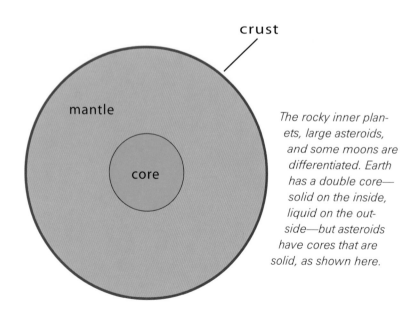

The rocky inner planets, large asteroids, and some moons are differentiated. Earth has a double core—solid on the inside, liquid on the outside—but asteroids have cores that are solid, as shown here.

common on Earth. However, there are minerals in meteorites that are not found on Earth. If you find a rock with those minerals, you definitely have a rock from space. Unfortunately, you can't tell just by looking. Rocks must be tested in a lab before you can be sure what minerals are in them. Scientists cut and grind meteorites into 0.03 millimeter thin sections mounted on glass microscope slides. They are thin enough to see through. By shining filtered light through them and studying the colors with a microscope, scientists can identify the minerals.

Augite (left) and diopside (right) are pyroxene minerals. These samples are from Earth, but the minerals also occur in meteorites.

A meteorite made of iron-nickel attracts a magnet.

Gem-quality olivine from Earth is called peridot.

TYPES OF METEORITES

There are three basic types of meteorites: stony meteorites, iron meteorites, and stony-iron meteorites.

Stony meteorites may be rocks from the mantles and crusts of differentiated asteroids or planets. Or they may be pieces of unmelted asteroids, accretions of early solar system dust.

A fresh and unweathered stony meteorite that fell in November 2008 at Buzzard Coulee, near the village of Marsden in Saskatchewan, Canada. The broken corner shows the lighter interior of the blocky stone. The cube measures 1 centimeter. —Photo courtesy Svend Buhl, www.meteorite-recon.com

Iron meteorites must have been part of the core of a growing planet or asteroid at one time.

This spectacular iron meteorite from Campo del Cielo, Argentina, is just one individual from a huge meteorite shower that produced some of the biggest iron meteorites on Earth. Cube for scale. —Photograph by Suzanne Morrison ©

Stony-iron meteorites contain roughly equal proportions of rock and metal. There are two kinds with completely different histories. They are rare and unique. You won't believe how beautiful some of them are.

Stony-iron meteorite slice from Glorieta Mountain, New Mexico. Cube for scale. —Courtesy Anne Black, www. impactika.com

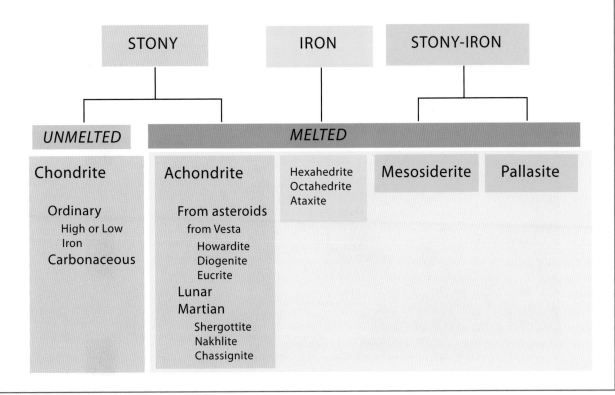

STONY	IRON	STONY-IRON

UNMELTED	MELTED		
Chondrite	Achondrite	Hexahedrite Octahedrite Ataxite	Mesosiderite Pallasite
Ordinary High or Low Iron Carbonaceous	From asteroids from Vesta Howardite Diogenite Eucrite Lunar Martian Shergottite Nakhlite Chassignite		

STONY METEORITES

Most meteorites are stony—rocks, in other words. The ones called ordinary chondrites (like those that fell on Park Forest) make up about 85 percent of all known meteorites. If you are lucky enough to find a meteorite, odds are it will be an ordinary chondrite. Despite their "ordinary" name tag, they are quite extraordinary compared to Earth rocks. Ordinary chondrites contain traces of the early solar system!

There are many kinds of chondrites, but as the name suggests, they all contain one essential ingredient: chondrules. The word comes from *chondros*, Greek for "grain." Chondrules are little spheres averaging about 1 millimeter across. The fact that they are spherical shows they were once liquid, because in space a liquid always takes a spherical shape. Astronauts on the space station drink liquids through a straw from a closed container because loose liquids float around in blobs instead of flowing downward and staying in a glass, as they normally would on Earth. Scientists think chondrules formed when small clumps of dust in the solar disk melted and cooled quickly, hardening the spherical blobs. How the melting took place isn't clear. But it certainly did, and more than once. Chondrites are filled with these little balls. Today, we think chondrules are some of the oldest materials in the solar system and that everything

This chondrule is unusually large: 7 millimeters across. Most are much smaller. Another one has fallen out of this meteorite (upper center).

was made from them. Most were melted as the planets formed, but some remained in the asteroids.

The parent bodies of chondrite meteorites were never completely melted and differentiated—if they had been, the chondrules wouldn't still exist. But they usually experienced some heat from radioactive isotopes that were trapped during the accretion process. This changed both their appearance and their chemistry. The meteorites that come from these parent bodies are classified by texture into categories that reflect

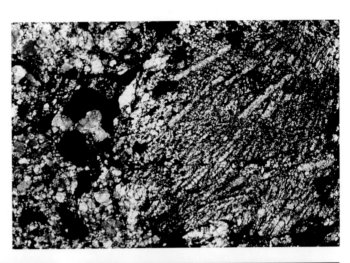

A photomicrograph (40x magnification) of a thin section of the Park Forest L5 chondrite, a type 5 chondrite, showing an altered radial pyroxene chondrule (the entire right side of the picture) with the surrounding material (left side). The chondrule has been damaged but not completely destroyed. Compare this to the unaltered radial pyroxene chondrule at below right. —Photo courtesy © Paul P. Sipiera

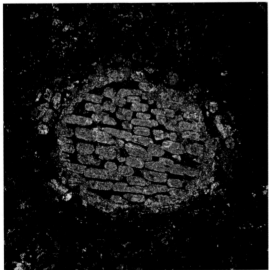

A typical barred olivine chondrule with parallel barlike plates of olivine. —Photo courtesy John Kashuba

A radial pyroxene chondrule that grew from a single point, viewed in thin section through a microscope.

the amount of heating. Types 1 and 2 weren't melted at all but were altered by water. Type 3 is considered unaltered. In types 4 and 5, the chondrules are less distinct. In type 6, melting was more extensive and the chondrules are almost gone.

Besides displaying chondrules in some form, ordinary chondrites typically contain small grains of silvery metal. As much as 23 percent of the iron in stony meteorites

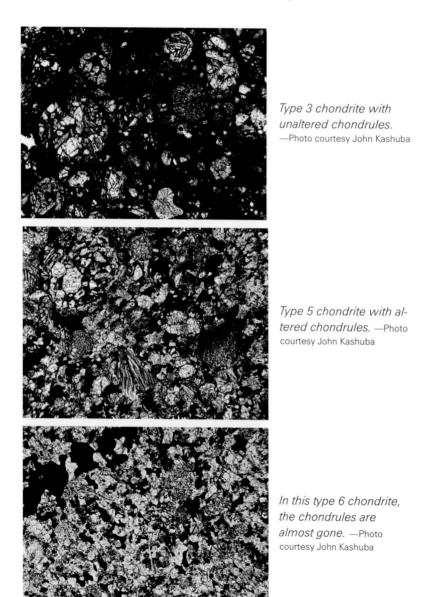

Type 3 chondrite with unaltered chondrules.
—Photo courtesy John Kashuba

Type 5 chondrite with altered chondrules. —Photo courtesy John Kashuba

In this type 6 chondrite, the chondrules are almost gone. —Photo courtesy John Kashuba can be in this elemental state, always mixed with nickel. Additional iron is found in combined forms as oxides, sulfides, carbides, and phosphides. Ordinary chondrites are further classified as high iron (H) or low iron (L or LL), depending on how much iron they contain. If you can see grains of metal scattered across the cut face of a stone, it's probably an ordinary chondrite. And like regular iron on Earth, iron in meteorites can rust.

What does this tell us about the classification of the Park Forest meteorite? It's an L5—low total iron, melted and therefore altered enough that it shows few chondrules. The Park Forest meteorite is also a breccia, meaning the source meteoroid had been broken apart and reassembled. Sometimes the internal pieces are sharply angular. Often the broken pieces appear against a background that is darker or lighter, indicating that rocks from two different asteroids were smashed together.

A slice of an ordinary chondrite called Sahara 98175 showing both metal grains (bright, light color) and chondrules (circular features). —Photo courtesy John Kashuba

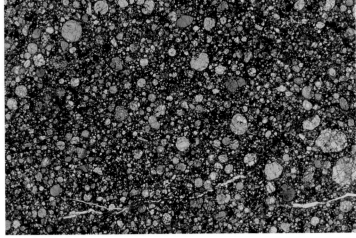

Chondrules, the first building blocks of the solar system, are easy to see in a type 3 chondrite.

Carbonaceous Chondrites

Carbonaceous chondrite meteorites are generally very dark, like the C-type asteroids they probably come from. Many things made of carbon, such as charcoal, are, after all, black. Often carbonaceous chondrites are full of colorful chondrules and some other unexpected things. Scientists think they are the most primitive of all meteorites, having formed early in the development of the solar system and changing little in their long history. In fact, some of them have a composition very similar to that of the Sun. These little lumps that resemble coal or charcoal are actually made of the same elements as the Sun but without the gases. They also contain flaky calcium aluminum inclusions that are very ancient and have never melted. Those particles are among the oldest solids in the solar system. Even more exciting, some of these rocks contain complex molecules called amino acids. All living things on Earth do too. Amino acids

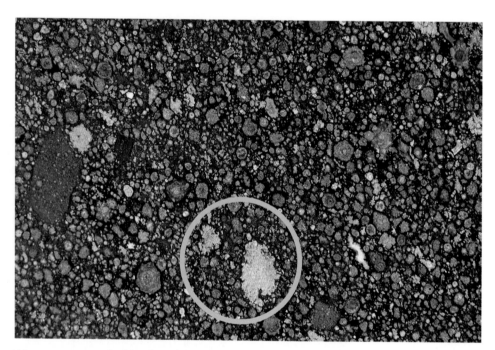

What can you see in this slice of a carbonaceous chondrite? There are colorful chondrules (the little spheres throughout) and calcium aluminum inclusions (two large white ones inside the yellow circle). Can you find a chunk of a different gray material on the far left? It's filled with even smaller chondrules. It may be the result of a collision with another parent body. —Photograph by Geoff Notkin/Aerolite.org, ©The Oscar E. Monnig Meteorite Gallery

combine to make proteins, which have many important functions in just about every living thing. It seems some of the basic ingredients of life formed in space long before Earth even existed.

Carbonaceous chondrites are very fragile. Unless they are recovered quickly or land in a very dry desert, rain and ice destroy them before they are ever found. They used to be exceedingly rare in meteorite collections, but that changed when two major meteorite falls occurred in 1969. Shortly after midnight on February 8, a brilliant fireball appeared in the sky over the little Mexican town of Pueblito de Allende in the state of Chihuahua. Explosions high in the air produced a shower of stones across an area of more than 180 square miles, the largest from any meteorite fall anywhere. One piece weighing 240 pounds shattered on impact, and more than 2 tons of carbonaceous chondrites were recovered in the first few weeks. Decades later, meteorites were still being found.

Two typical small carbonaceous chondrites from the Murchison fall. They are very fragile and break easily.

WATER, WATER EVERYWHERE

You've seen the beautiful pictures of Earth taken from space: the blue marble, covered with more water than land. And you've probably seen pictures of other planets—barren, lifeless, and rocky, with no water at all. So you may find it surprising that water is actually quite common in the solar system, and that it was abundant in the earliest days when planets and asteroids were forming. Recently, water has been found in the form of ice on Mars and the Moon, and we know there is ice on the outer planets and their moons. In a few cases, actual drops of incredibly ancient water have been found inside salt crystals in meteorites. How did they get there? That's still a mystery.

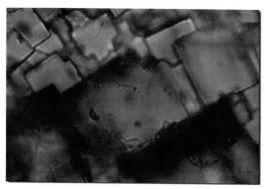

Tiny pockets of water in a purple salt crystal found in the Monahans meteorite, a chondrite that fell in Texas in 1998. Image size is 1 millimeter across. —Photo courtesy NASA

This photo of Earth was taken by Apollo 17 astronauts in 1972.
—Photo courtesy NASA

WHAT'S IN A METEORITE NAME?

Meteorites are generally named for a nearby geological feature or the nearest physical location with a post office, like Murchison and Allende. Meteorites from Antarctica and Northwest Africa use geographical regions and abbreviations.

Meteorite types are named for people, such as howardites, named for Edward C. Howard, a chemist; for minerals, such as enstatite chondrites, named after enstatite, a pyroxene mineral; or for places, like aubrites, named for Aubres, France. The names are usually based on the first specimen of a new type that's described. In cases where a large meteorite breaks into many smaller pieces, as happened at Park Forest, all the pieces have the same name.

On September 28, 1969, a fall of a different type of carbonaceous chondrite occurred in Australia. It was Sunday morning and most people in Murchison, Victoria, were on their way to church when a bright fireball dropped stones over a 5-square-mile area. People quickly gathered the meteorites, more than 220 pounds. It was fortunate that the fall happened over a populated area (and no one was hurt) because these rocks are very delicate and would have disintegrated quickly if left out in the weather. The Murchison meteorites turned out to be very important to researchers. They contain water (combined in minerals) and amino acids, as well as microscopic interstellar particles and in some cases even exceedingly tiny diamonds, which can form only under great pressure in supernova explosions.

Achondrites

Chondrites are the most common stony meteorites, but there are also achondrites. The prefix *a* means "not" or "none," so achondrites do not have chondrules. These are meteorites from asteroids, planets, or our Moon, which have become entirely differentiated due to complete melting. There are no rocks with chondrules on Earth either. They disappeared long ago, as the early Earth was melting.

Achondrites from Asteroids. Vesta, the fourth asteroid discovered, is also one of the largest. With a diameter of about 330 miles, it can sometimes be seen without a telescope. Spectroscopic studies reveal that, as it rotates, its surface composition looks different in different areas. This indicates that Vesta isn't homogenous. In some areas it is made of basalt—an igneous rock formed by volcanoes erupting onto the surface. Other areas are igneous but made of rocks that solidified beneath the surface. Still other areas are cratered. Erupting volcanoes on such a little world? It seems impossible, but this asteroid clearly has a violent history. It has one enormous crater and isn't quite round. Big chunks seem to be missing. We now know that pieces of Vesta are resting in museums and meteorite collections on Earth.

Because about 6 percent of all meteorites on Earth seemed to be related to Vesta, there was a lot of interest in studying this asteroid up close. Meteoriticists wanted

The logo on the Dawn *mission patch shows the spacecraft approaching Vesta and Ceres. Vesta's large crater at its south pole has a central mountain almost thirteen miles high, over twice as tall as Mount Everest. The crater is surprisingly young at just 1 billion years.* —Photo courtesy NASA

to know if the big crater was the source of three groups of meteorites that are very different in some ways, but chemically similar: howardites, eucrites, and diogenites. They wanted to know if the family of asteroids called Vestoids, named because their reflectance spectra resembles that of Vesta, originally came from there. Some of them are in the asteroid belt, but others have become near-Earth asteroids, which would explain how related meteorites ended up here.

On September 27, 2007, NASA launched the *Dawn* spacecraft on a historic mission to visit two of the largest residents of the asteroid belt—Vesta and Ceres. *Dawn* arrived at Vesta on July 16, 2011, and settled in for a one-year orbit of the asteroid before continuing on to Ceres. It confirmed that Vesta is differentiated and has an iron core, and that the meteorites come from there.

Dawn made this picture of Vesta's south polar region on July 9, 2011, from a distance of 26,000 miles, just before beginning its one-year orbit. —Photo courtesy NASA/JPL-Caltech/ UCLA/MPS/DLR/IDA

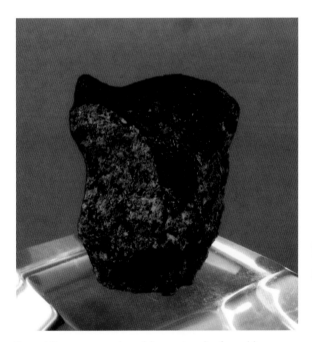

Camel Donga, an achondrite meteorite found in Australia in 1984, came from Vesta. It has a very shiny black crust, typical of eucrites, one of the three kinds of meteorites now known to be from Vesta. —Photograph by Suzanne Morrison © Aerolite Meteorites, www.aerolite.org

Achondrites from the Moon. Look at the Moon and what do you see? Craters and more craters. For the longest time people thought the Moon was the source of all meteorites. (They also thought the craters were volcanoes, not caused by meteorite impacts, as we now know they are.) Unfortunately for the lunar origin theory, meteorite orbits, calculated from observations of fireballs, pointed to the asteroid belt, not the Moon. Then in 1981, Japanese scientists working in Antarctica realized they had found a meteorite from the Moon. The rock, a breccia of angular fragments cemented together, was so similar to samples brought back from the Moon by Apollo astronauts that there could be no doubt. Laboratory studies of the meteorite's mineral and isotope compositions confirmed it. Today about seventy-seven lunar meteorites have been found. They are especially helpful to our understanding of the Moon because astronauts explored only a small area of the Moon, whereas lunar meteorites represent a much wider sampling, coming from the highlands (light areas) and lowlands (dark areas). Most lunar meteorites are breccias, and all are more than 3 billion years old.

This picture of the Moon was taken from the command module by Apollo 12 *astronauts after leaving lunar orbit on November 24, 1969.*
—Photo courtesy NASA

ALHA 81005, the first recognized lunar meteorite, is a breccia from the highlands. It is similar to rocks collected on the Moon. (ALHA stands for Allan Hills, the region of Antarctica where it was found.) The cube measures 1 centimeter. —Photo courtesy NASA

A lunar breccia collected on the rim of the Moon's North Ray Crater in April 1972 by astronaut Charles M. Duke Jr., Apollo 16 lunar module pilot.
—Photo courtesy NASA

DaG 400, found in the Libyan Sahara desert, is a meteorite from the lunar highlands. Some scientists think this may be a rock from the far side of the Moon, the side we never see from Earth. (DaG stands for Dar al Gani, the region where it was found). —Photograph by Geoff Notkin/Aerolite.org, ©The Oscar E. Monnig Meteorite Gallery

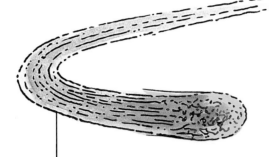

DATING ROCKS AND RADIOACTIVE CLOCKS

Scientists can tell how old meteorites and Earth rocks are by using their built-in clocks—radioactive isotopes that decay over time. Scientists determine the age of a rock by comparing the amount of decay products, which form at a known rate, with the stable parent isotopes present in a sample. This method doesn't necessarily reveal the absolute age of a rock, but it does reveal how much time has passed since the rock or its minerals melted. In the case of unmelted meteorites, this can be the time since they first accreted in the solar nebula. Similar tests can tell us how long a meteorite was in space and how long it has been on Earth. The presence of radioactive isotopes does not make meteorites radioactive. They are not dangerous.

Apollo 16 *lunar module* Orion *with lunar rover and astronaut John W. Young in 1972.* —Photo courtesy NASA

Achondrites from Mars. We haven't been able to bring rocks back from Mars yet, but in the 1980s geologists discovered there were samples here already. While studying air trapped in an Antarctic meteorite, they found that the gases matched the atmosphere on Mars as measured by the Viking spacecraft that landed there in 1976. Like their very distant lunar cousins, rocks from Mars were blasted off the surface in an energetic asteroid impact and then made their way to Earth. Compared to the more-than-3-billion-year-old lunar rocks, meteorites from Mars are surprisingly young. Some are a mere 180 million years old, and only one is older than 1.3 billion years.

With the exception of one very old meteorite, all martian rocks fit into three groups, each named for the place where the first one in that group was found and identified. They are called shergottites (from Shergotty, India), nakhlites (from El-Nakhla, Egypt) and chassignites (from Chassigny, France). They differ from each other in appearance, but comparisons of their mineral and isotope compositions prove they all come from Mars.

About ninety Martian meteorites have been found at the time of this writing. This does not mean they come from ninety different places on Mars, however. They are probably the result of a few large impacts on Mars. Surprisingly, all are igneous rocks. None are breccias like the rocks from the Moon.

A shergottite from Mars, the Zagami meteorite fell in Nigeria in October 1962. Only one stone was found. This piece of the stone measures about 6 inches across. Without the shiny black fusion crust, found only on meteorites, you would never know it's not a rock from Earth.

METEORITE ALH 84001:
EVIDENCE FOR ANCIENT LIFE ON MARS

ALH 84001 is a controversial, one-of-a-kind Martian meteorite. The cube measures 1 centimeter. —Photo courtesy NASA

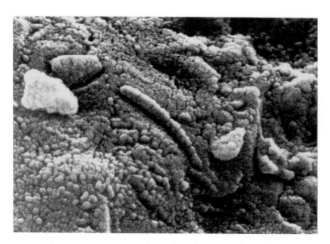

These tiny structures may be microbacteria, less than one-thousandth the size of Earth bacteria. They were found inside the carbonate deposits in ALH 84001. —High resolution electron microscope image courtesy NASA

The most famous meteorite in the world—one that may contain evidence of life on Mars—is a rock found in Antarctica in 1984: ALH 84001. (Like ALHA, ALH stands for Allan Hills, in Antarctica.) Tests prove it's related to the other meteorites from Mars, but it's otherwise unique and much older—4.09 billion years. It's an igneous rock, and although it formed deep inside Mars, it was shocked many times by impacts. At some point water seeped through cracks and deposited tiny beads of carbonate minerals in the rock. Later, an asteroid impact ejected it from Mars and catapulted it into the inner solar system. After 16 million years it plunged through Earth's atmosphere and landed in Antarctica, where it waited 13,000 years for someone to pick it up.

David McKay and Everett Gibson led a team of scientists who studied the strange rock, especially the carbonate deposits. They found compounds typical of decaying organic material on Earth and tiny wormlike structures that resemble bacteria. On August 7, 1996, they announced the discovery of possible fossil life on Mars. It's a dramatic discovery, but not everyone is convinced. Scientists continue to study the meteorite, but it will probably take a mission to Mars that can return samples of similar rocks to settle the question of whether life existed on ancient Mars.

MARS HAS METEORITES TOO

NASA's twin rovers—*Spirit* and *Opportunity*—were sent to Mars to search for signs of water on the Red Planet, but what do you suppose they found? Meteorites! Designed to function for only ninety days, the rovers continued to work for more than six years after landing in January 2004. They amassed a huge amount of data, traveled more than 20 miles, and sent home more than 250,000 pictures. In addition to cameras, the rovers carried equipment, including a spectrometer and magnets, for sampling soil and rocks that scientists at the Jet Propulsion Laboratory in Pasadena, California, found interesting. Using an arm with an elbow and wrist, the rovers could get very close to rocks of interest. A device called the Rock Abrasion Tool would then grind the rock surface to expose the inside. Using these tools, the rovers discovered five meteorites on the surface of Mars. They're not meteorites *from* Mars, they're iron chunks from somewhere in the asteroid belt that fell onto Mars in the distant past. Because Mars is very dry and its atmosphere has little oxygen, iron doesn't rust there like it does on Earth.

In recognition of the rovers' spectacular success, two asteroids have been named for them: 37452 Spirit and 39382 Opportunity.

Spirit *discovered this iron meteorite called Allan Hills, after a region in Antarctica where many meteorites have been found, including several from Mars.* —Photo courtesy NASA/JPL-Caltech/Cornell

Contact with the rover Spirit *was lost after its last communication on March 22, 2010.* —Exploration Rover Mission, courtesy NASA/JPL/Cornell

IRON METEORITES

Some meteorites are made entirely of iron and nickel. Iron comes from the cores of differentiated worlds such as rocky planets and spherical asteroids. M-type (metallic or metallic and rocky) asteroids seem to be mostly iron. What does it take to expose an iron core? Did some asteroids lose their mantle and crustal rocks in unimaginably huge impacts—impacts so enormous they were fortunate to survive at all? It looks like it. We know there were many massive impacts in the early years of the solar system. The Moon and inner planets are covered with craters. Only Earth, with its atmosphere and active mountain building and volcanism, has lost most of its scars

Tim Heitz, a meteorite collector and dealer, really, really loves meteorites. Here he sits atop the 37-ton Campo del Cielo meteorite, known as El Chaco, in Argentina, the second-largest iron meteorite known on Earth. —Photo courtesy Tim Heitz of Midwest

Hoba, the largest iron meteorite in the world, weighs 60 tons. It has never been moved from the place where it was found, in Namibia. A small amphitheater has been built around it.
—Photo courtesy David Mouat, Desert Research Institute, Nevada System of Higher Educaton

from that early period. The strange shapes of some metallic asteroids even suggest the cores were molten—melted—when collisions occurred. Remember the giant dog bone on page 21?

The traditional classification system for iron meteorites is based on how much nickel they contain and how they look after they are sliced and etched with acid, a process that reveals their internal structure. Hexahedrites have less than 6 percent nickel, octahedrites have 6 to 13 percent nickel, and ataxites have more than 13 percent nickel. As you might expect, these meteorites are attracted to magnets.

Cutting up an iron meteorite isn't easy. It takes special equipment and often destroys a saw blade in the process. But it's worth doing, partly to reveal the crystal shapes, which are unlike anything found near the surface of Earth. Material from Earth's core and core-mantle boundary may look a lot like meteoritic iron.

The amount of nickel in the iron when it's molten determines which minerals form as it cools and what the final crystal structure looks like. The two most important iron-nickel minerals in iron meteorites are called taenite and kamacite. High-nickel ataxites are pure taenite. Low-nickel hexahedrites are pure kamacite. Octahedrites, the most common iron meteorites, are a mixture of both, and the intergrowth of crystals can make a very beautiful pattern when polished and etched. It's called a Widmanstätten pattern, after Count Alois Beckh von Widmanstätten, who usually gets credit for discovering it. However, William Thomson actually discovered it first, and his name would be easier to pronounce.

The surface of this 66-pound iron meteorite from the Henbury crater field in Australia was melting and flowing while it traveled through the atmosphere.

This acid-etched iron meteorite slice (9 inches wide) is from Lake Murray, Oklahoma. It is an octahedrite and clearly shows the Widmanstätten crystal pattern.

A superb example of a shrapnel meteorite, a piece of iron that has sharp edges because the meteoroid exploded violently as it moved through Earth's atmosphere, from the Kamil Crater, Egypt. Cube for scale. —Photograph by Suzanne Morrison © Aerolite Meteorites, www.aerolite.org

STONY-IRON METEORITES

Stony-iron meteorites are a mix of roughly equal proportions of rock and metal. There are two completely unrelated kinds.

Mesosiderites (from the Greek words *mesos*, meaning "half," and *sideros*, meaning "iron") were probably created by energetic impacts like those that formed the metallic asteroids. A tremendous collision in which an iron asteroid crashed into the crust of a differentiated asteroid may have broken and melted them, resulting in a new asteroid that's a jumbled mixture of both.

Pallasites—named not for the asteroid Pallas but for Peter Simon Pallas, who described one of these meteorites discovered in Siberia in 1749—were probably also created by large impacts, but they are not pieces from multiple parent bodies. They seem to be samples from the core-mantle boundary regions of individual differentiated asteroids. When the core and surrounding mantle material was molten, the rocky minerals separated from the heavier iron and nickel, forming beautiful crystals of the mineral olivine surrounded by a network of iron. Or they may be the result of an impact that melted core and mantle material together. They are truly spectacular meteorites, unlike any Earth rock you will find.

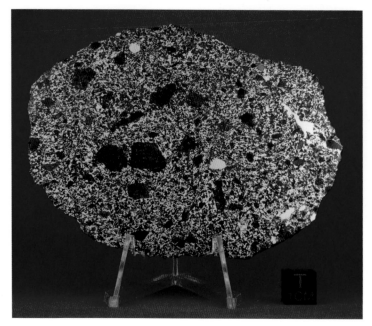

This slice of the Vaca Muerta mesosiderite from the Atacama Desert in Chile, shows a mix of silver-colored iron and dark gray stony material. —Photograph by Suzanne Morrison © Aerolite Meteorites, www.aerolite.org

An etched slice of the Seymchan pallasite from Russia. Note how etching the iron with acid brings out Widmanstätten patterns even in a pallasite stony-iron meteorite. Cube for scale. —Photo courtesy Anne Black, impactika.com

A slice of a pallasite from Esquel, Argentina. It's thin enough to allow light to shine through the yellow olivine crystals. —Photo courtesy Anne Black, impactika.com

The Peekskill, New York, multiple fireball shortly after it fragmented on October 9, 1992.
—Photo courtesy *Altoona Mirror*

What Happens When Asteroids Strike?

"We expect an asteroidal object 1 meter [3.3 feet] in diameter or larger to strike Earth's atmosphere about forty times per year," says Donald Yeomans, manager of NASA's Near-Earth Object Program at the Jet Propulsion Laboratory in Pasadena, California. Few are seen, however, because most of Earth is covered by oceans and much of the land is rural or uninhabited. When an asteroid enters the atmosphere, a visible fireball passes through the sky. If it's big enough, the force of the impact may blast out a large crater.

FIREBALLS

If you are very lucky, you might see a bright fireball. You might even see it break up, sometimes into many pieces, and continue across the sky in a dazzling display that resembles Fourth of July fireworks. You might hear sonic booms or other sounds. It's a breathtaking experience. When the light dies out, you might even see rocks falling from the sky. At times like this, it's hard to think about anything. But knowing what to look for can help you find meteorites later—if they aren't just falling through your roof like the ones in Park Forest.

A big fireball is nothing like the meteors that flash into existence and burn up high in the atmosphere during a meteor shower. Fireballs are brighter, bigger, totally unexpected, and sometimes noisy. Daytime fireballs can look brighter than the Sun. And even though they may seem close, they're probably a lot higher than you think.

Fireballs become visible about 60 miles above the ground. And their light tends to go out, or be extinguished, when they are still quite high. That's what makes it so difficult to find any meteorites that make it to the ground. It's hard to tell exactly where something will land when you only see it while it's high up in the sky. Meteorites from a fireball that seemed to pass "just over that hill" often land in the next state.

People witnessing a bright fireball from different places tend to judge its location differently. To find a meteorite on the ground from the accounts of witnesses, you must have reports from several places and plot the sightings on a map to get an idea of where it actually was when its light was extinguished. When a meteoroid loses its velocity and stops glowing, it begins to fall by gravity alone. Any meteorite that makes it to the ground usually lands within a few miles of where it stopped glowing.

The Great Meteor of August 18, 1783, seen in Newark-on-Trent, England. A single huge fireball blazed across a clear evening sky and broke into many pieces. It was seen over a wide area, but no stones were recovered.
—Engraving by Henry Robinson

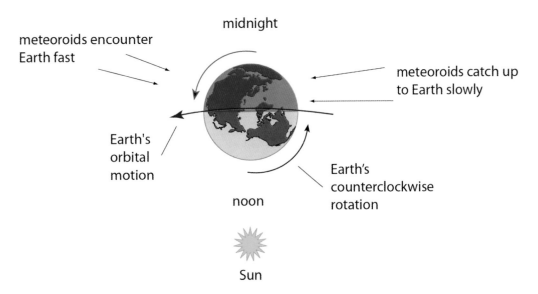

Meteoroids approaching Earth from the front (as it travels in its orbit) are hit faster than meteoroids that have to catch up to Earth from behind.

A meteoroid enters Earth's atmosphere traveling between 10 and 26 miles *per second*. That's really fast: 36,000 to 94,000 miles per hour. The actual speed depends on its mass and also on the direction and angle it's coming from—whether it's approaching Earth head-on or catching up to Earth, which travels in its orbit around the sun at 18 miles per second.

At about 60 miles above the ground, the atmosphere becomes dense enough to create some drag, slowing meteoroids and beginning to heat them. When the outer surface of a meteoroid reaches about 3,000 degrees Fahrenheit, it melts and begins to glow. The material that's sloughed off the surface is vaporized. As heating continues, the air around the rock becomes ionized, with atmospheric atoms and molecules losing electrons. As the electrons are recaptured by the atmosphere, they release light, making the air glow as well. The luminous shell around an incoming meteoroid can be huge compared to the size of the rock it surrounds. A glowing mass of air hundreds of feet across—the fireball you see—might contain a rock only 1 foot in diameter.

Besides seeing a dramatic light show, people often hear sounds like thunder, produced by a shock wave in the atmosphere. Sometimes it sounds more like a jet plane breaking the sound barrier. A meteoroid on its way to the ground may make a "whomp whomp" sound that reminds people of a helicopter or a car driving on a flat tire. The strangest noises fireballs make are like hissing radio static. A huge fireball can actually create low-frequency radio transmissions that are converted to sound.

Ablation Is Only Skin Deep

As the incoming rock loses material from its surface, a process called ablation, it gets smaller. The lost vaporized material cools and condenses, creating the train of dust often seen behind a fireball. How much material is lost depends on the meteoroid's

Russian artist P. J. Medvedev made this painting after witnessing the Sikhote-Alin fireball in 1947. The smoke train contained more than 200 tons of sloughed-off material. Ten years later the Soviet Union issued a postage stamp commemorating the event.

This beautiful cone-shaped meteorite from Oman is covered with flow lines. You can see that only the outer surface was melting on its very brief trip through the atmosphere. The cube measures 1 centimeter. —Photo courtesy Svend Buhl, www.meteorite-recon.com

composition, size, and velocity. A rocky meteoroid will probably lose more material than one made of iron, but even iron meteoroids can lose most of their mass. The force of air pressure can be enormous and often causes meteoroids to break up in flight. Most meteorite showers, in which hundreds or even thousands of stones fall over a wide area, result from the shattering of a single mass.

Please don't get the idea that the whole rock is melted—literally molten—when it hits the ground. This is never the case. Space is very cold—almost minus 460 degrees Fahrenheit, an unimaginable temperature—and most meteoroids in our solar system have been out there for 4.6 billion years. They are very cold when they enter Earth's atmosphere, and the time they spend traveling through the atmosphere is brief. Only the outer surface of the rock melts. Sometimes a freshly fallen meteorite is warm, but it's not unusual to find stones that are quite cold, even frosty.

Meteorites that manage to make it to the ground probably do not bear much resemblance to their earlier selves in space. No one has ever seen one up close before it entered the atmosphere, but we know it's a harrowing trip, and a lot can happen.

Meteoroids heat up, lose mass, and can break up explosively. When the surface is melting, it can flow like a liquid. Flow lines are a sure sign that a rock is a meteorite. Molten material can even flow to the back of the meteorite, creating a distinct rollover lip. The shape a meteorite is in when it finally arrives also depends on how it was oriented on its journey. Meteoroids that tumble randomly may become quite blocky or rounded on all sides. If a meteoroid manages to stabilize itself, it can develop a shield or cone shape.

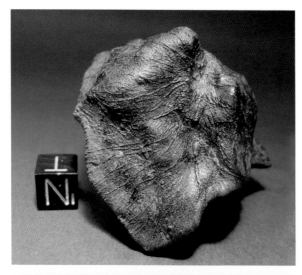

Flow lines are obvious on this meteorite from Australia, known as Millbillillie. The cube measures 1 centimenter.
—Photo courtesy Svend Buhl, www. meteorite-recon.com

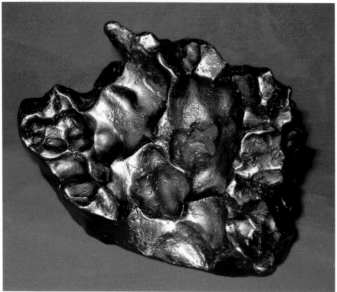

You can really see the "thumb-prints" in this Sikhote-Alin iron meteorite from Siberia.

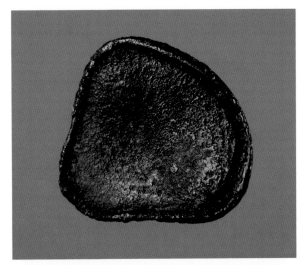

This Sikhote-Alin iron meteorite has formed a perfect rollover lip, something that happens only rarely, during its brief flight through the atmosphere. —Photograph by Suzanne Morrison © Aerolite Meteorites, www. aerolite.org

A collection of very fresh meteorites that fell at Bensour, Algeria, on February 11, 2002. They were picked up immediately and display excellent examples of the fusion crust on fresh meteorites. They are type 6 chondrites, with very few chondrules visible.

Iron meteorites often have deep pits, called "thumbprints," that are a sure sign of ablation. Occasionally they develop holes that may go all the way through. Sometimes iron meteoroids explode violently, and the individual meteorites that result are distorted and have sharp edges. They are called shrapnel meteorites because they are like pieces of exploded bombs.

A fresh, unweathered stony meteorite will be covered by a fusion crust, a thin outer layer (less than 1 millimeter thick) created by melting of the stony material. It's glassy, usually quite black, and generally dull. Irons also form crusts, but they are very thin and rarely survive weathering for long.

The Strewn Field

Meteorite showers can produce many meteorites—sometimes hundreds or more. If you could record the location of each one, you would see that they don't just drop in a jumble. Meteorites from the same shower fall in a predictable pattern, and understanding that pattern is helpful in trying to figure out their distribution.

Even after the main mass breaks up, the pieces continue to travel together, and they tend to fall in a generally elliptical, or oval, pattern called a strewn field. Momentum carries the larger pieces farther within this area. If you can find the long axis of the ellipse, you can determine the direction the meteorites were traveling.

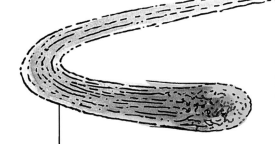

FIREBALL OBSERVER'S CHECKLIST

If you see a fireball, it's worthwhile to record as much information about the event as possible. This information may help you or others find the related meteorites:

1. Date and time
2. Your longitude and latitude (use a GPS device or look it up at the library or online)
3. Reference landmarks (for example, just over the neighbor's house)
4. What direction was it going?
5. How long did it last?
6. If it's nighttime, can you identify the constellations it went through?
7. How bright was it? Did it cast a shadow?
8. How large was it?
9. Did it leave a smoke train?
10. Did you hear any sounds?
11. Did you see anything fall? Where?

If you did see something fall, answer as many of the following questions as possible:

12. Did it make a crater or small hole?
13. How long did it take you to find it?
14. Was it warm or cool?
15. How much does it weigh?
16. Before removing a meteorite, take photos of it where it landed and get the GPS coordinates if you can. Also take pictures of any hole or crater after removing the meteorite.

17. If the meteorite is very fresh, wear gloves and seal it in a plastic bag to avoid contaminating it.
18. Remember that a meteorite belongs to the owner of the land where it falls. If searching on private property, be sure to get permission first.
19. And most important, *don't forget to enjoy the show!* You may not find a meteorite, but the experience of seeing a fireball is a treasure in itself. It's something you'll never forget.

IMPACT CRATERS

Just as people once thought rocks couldn't fall from the sky, no one imagined that some of the largest craters and huge circular features on Earth could possibly be caused by impacts of objects from space. Scientists thought they were all volcanoes or sinkholes. Although the Moon is covered with craters, as are Mercury, Mars, and many moons of the outer planets, our home world is different. Because we have a thick atmosphere, weather, and a geologically active planet, we don't see many signs of impact. Old craters have been erased by geologic processes over millions of years. However, more than 170 impact craters have been identified on Earth.

Some craters are so huge you can't recognize them from the ground. They must be seen from space to be appreciated, and they are extremely old. No meteorites have been found around the largest ones. The impacting asteroids were completely vaporized, or their meteorites have weathered away.

Impact craters continue to be discovered using satellite images available on the Internet. In 2008, Italian researcher Vincenzo de Michele discovered a crater 147 feet wide in southern Egypt using Google Earth. The Kamil Crater may be only a few thousand years old, which would make it one of the youngest craters on Earth. It is well preserved and surrounded by iron meteorites.

The presence of so many craters around the world tells us something important. Our Earth was once like the Moon, a cratered world. The pattern of cratering we see on the Moon suggests something comforting, though. The large dark areas (called maria, plural of the Latin word *mare*, meaning "sea") are actually huge basins filled with basalt that overlie the older cratered surface. This means the basins are younger. We now know that an intense period of cratering in the early solar system, which certainly impacted Earth, ended nearly 4 billion years ago. Earth was once bombarded by asteroids, but things are quieter now, and that's reassuring.

However, we still get some fairly large visitors from time to time. The most recent arrived on February 12, 1947, in eastern Siberia in the Sikhote-Alin Mountains near the Sea of Japan. At about 10:30 a.m., a flaming fireball appeared in the cloudless sky. It was as bright as the Sun and cast shadows on the ground as it moved from north to south, trailing a huge smoke train. It passed near several towns and disappeared over a wooded area. Witnesses reported a tremendous explosion and the sound of rolling thunder. The shock wave was felt for hundreds of miles. In the deep snow of the Siberian winter, it took days to locate the point of impact. When searchers finally found it, they discovered over one hundred small craters. The largest was 85 feet wide. Tons of iron meteorites, some shrapnel and some with "thumbprint" pits from surface melting, were scattered in the snow and embedded in trees. Today fragments of the Sikhote-Alin iron meteorite are common and often show up on eBay.

Lake Manicoua-gan is Canada's second-largest crater. It's 62 miles in diameter and 214 million years old. —Space shuttle photo courtesy NASA

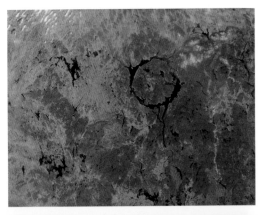

Aorounga Crater in Chad, in the Sahara Desert, is a heavily eroded impact crater. It's 8 miles in diameter and about 300 million years old. Wind-blown sand covers much of the crater. —Space shuttle photo courtesy NASA

The chunk of iron that crashed in the Sikhote-Alin Mountains probably had a mass of about 1,000 tons in space, but only about 70 tons made it to the ground. The event was huge, but it was much smaller than ancient impacts that have produced giant craters.

In recent years, several programs have been established by various nations to look for asteroids that might be on a collision course with Earth.

Meteor Crater

One of the most famous meteorite craters in the world is located in a remote area near Winslow, Arizona. It was created about 50,000 years ago during Pleistocene time. No people lived in North America back then, but plenty of animals did. Mammoths, horses, and giant sloths lived in this region, now known as the Colorado Plateau.

Perhaps the impact happened on a warm, sunny day. There would have been no warning. A small iron asteroid about 130 feet across entered Earth's atmosphere at a low angle. It had fractured at some time in the past and now about half of it broke into many pieces as it hurtled toward the ground. Thousands of small fragments were scattered on the plains to the west.

Impact crater expert H. Jay Melosh estimates that the remaining half hit the ground at a speed of 26,800 miles per hour. At that moment, a shock wave pushed ahead of the iron mass and down into the rocky plain, then back through the meteorite. The meteorite and impacted rock were subjected to enormous pressures, and a huge crater formed. The blast pulverized and melted rock, and also distorted pieces of the meteorite into shrapnel. An explosion equivalent to a 2.5-megaton bomb (2.5 million tons of TNT) raised a mushroom-shaped cloud above the plain and rained debris onto the crater rim. Originally 1,200 feet deep, the crater was halfway filled with rubble that settled back inside it after the explosion. Pieces of iron from the fractured mass and a layer of ejected rock were spread around the outside of the crater.

In addition to the shrapnel meteorites, people have found many magnificent iron specimens in the area, starting in the late 1800s. They are called Canyon Diablo irons after a winding canyon just to the west. So many meteorites were lying around that Daniel M. Barringer, a lawyer, geologist, and mining engineer from Philadelphia, filed a claim on the land in 1902. Barringer was convinced the crater was caused by the

Meteor Crater, in northern Arizona, is nearly 1 mile wide and 600 feet deep.
—Photo courtesy John D. Parker, Principal Photography

impact of a huge meteorite, and he wanted to mine the iron mass he assumed must be under the crater floor. He worked on it until his death in 1929 but never found the great iron body he sought. Most of the impacting asteroid had simply vaporized. Despite Barringer's efforts to publicize the crater's meteoritic origin, other scientists were not convinced.

It seems amazing that although the iron fragments were identified as genuine meteorites, the crater wasn't associated with them. Perhaps it was because of the crater's relatively close proximity to the volcanoes near Flagstaff, Arizona. Several noted geologists visited the site and pronounced it an old volcanic steam explosion. At the time, they couldn't accept what today seems so clear. Finally, in 1963 Eugene Shoemaker proved the crater was caused by a rock from space. He studied overturned layers of rock in the walls and analyzed rock beneath the floor. What he found was rock that had been deformed by the shock wave and melted by the heat of the blast with embedded meteoritic droplets and a lens-shaped area of breccia. Shocked rock is a sure sign of impact.

It's worth the effort it takes to visit Meteor Crater. Standing on the rim, you can't help but wonder, *What if it happened today?*

This 15-pound Canyon Diablo iron was found long ago near Meteor Crater, Arizona. Cube for scale. —Photograph by Suzanne Morrison © Aerolite Meteorites, www.aerolite.org

DEATH BY METEORITE?

Let's face it. Even with the atmosphere and ablation slowing meteorites down, they're traveling about 200 miles per hour when they hit the ground. There's no record of anyone being killed by a meteorite in modern times, but a few people have been hit. Ann E. Hodges of Sylacauga, Alabama, was badly bruised in 1954 when a meteorite crashed through her roof and bounced around, damaging a radio before striking her. In 1992, a boy in Uganda was struck by a small stone that first hit a tree. A cow was killed in Valera, Venezuela, in 1972. Accounts of a dog dying in the Nakhla, Egypt, fall in 1911 can't be confirmed.

A collision with a large asteroid would definitely cause death and destruction. We do know there are plenty of asteroids that come by regularly. Every month about one thousand asteroidal chunks of various sizes come closer to Earth than our Moon, a distance of 238,857 miles. One of the closest encounters we know of happened on February 4, 2011, when a small Apollo asteroid passed by only 3,405 miles above the Pacific Ocean. Whether Earth's gravity can pull an asteroid into our atmosphere depends on the asteroid's mass, velocity, and angle of approach.

Besides the near-Earth asteroids and near-Earth objects, astronomers maintain a list of PHOs—potentially hazardous objects, both asteroids and comets. Surveys in several countries continue to identify and track these objects, but often they are discovered only hours before their closest approach. On October 6, 2008, astronomer Richard Kowalski, a member of the Catalina Sky Survey in Tucson, Arizona, discovered a small asteroid on a collision course with northeastern Sudan. Only twenty hours later, asteroid 2008 TC3 exploded in the air above the desert in the predicted location. Scientists and students rushed to the site and recovered many meteorites. The fall was named Almahata Sitta for a nearby train station.

Eighteen-year-old Michele Knapp didn't have far to look for the Peekskill, New York, meteorite on October 9, 1992. It hit the rear end of her 1980 Chevy Malibu. The following year, the famous car went on a world tour.

TEKTITES

Tektites, curious pieces of glass people often confuse with meteorites, look aerodynamic—like they were flying through the air when they were molten. In fact, this is exactly what happened. Tektites (from the Greek *tektos*, meaning "melted") were created by enormous impacts, larger than any in recorded history. They are pieces of rock torn from Earth's surface and blasted hundreds of miles into space. Melted first by the impact, they melted again when they fell back to Earth. Most tektites are black and resemble obsidian, a volcanic glass, but they are chemically different. Beautiful green moldavites from the Czech Republic, named for the Moldau River, originated in the Nördlinger Ries Crater in Germany. Four major areas of tektites are known worldwide. The largest, stretching from Australia to Southeast Asia, is not associated with any known crater. These tektites are thought to be about 700,000 years old.

Moldavites, beautiful green tektites from the Czech Republic, are often deeply grooved and pitted. Sometimes they are made into jewelry. —Photo courtesy John Kashuba

Typical aerodynamic shapes of tektites from Southeast Asia are dumbbell, flattened disk, and teardrop. —Photo courtesy of The Meteorite Exchange, Inc.

MASS EXTINCTIONS FROM ASTEROID IMPACTS

Many scientists think the dinosaurs went extinct 65 million years ago when an asteroid about 6 miles across crashed into the ocean in the vicinity of what is now the Yucatán Peninsula in Mexico. However, there is still some disagreement about it. Some paleontologists, scientists who study prehistoric life, maintain that dinosaurs were actually dying out for a long time before the catastrophic event. Others say some species of dinosaurs survived for a while afterward. But the fact is, they're gone, and we're here—and those two realities are related. If the dinosaurs were around today, mammals (like us) probably would still be mouse-sized little creatures sneaking around at night. The extinction of the dinosaurs opened up new evolutionary niches for mammals to fill.

It wasn't just the dinosaurs that were wiped out 65 million years ago. About 75 percent of all living species, on land and in the sea, disappeared. Mass extinctions like this have happened at least four other times in Earth's history. The largest extinction, an event called the Great Dying, occurred about 250 million years ago. An unbelievable 70 percent of all land species and 96 percent of all marine species were destroyed. It took 30 million years for vertebrates to recover. Perhaps an asteroid or comet was to blame. We don't know.

It's a carbonaceous chondrite!

How Can You Find Meteorites?

Now that you know something about meteorites, you're probably anxious to get out and start finding them yourself. If rovers on Mars found five in just 20 miles of travel, surely a person could find one here on Earth. It isn't as easy as you might think, though. It's true that they are different from Earth rocks, some completely so. But they are still rocks—and Earth has a lot of rocks. Examining every one to see if it's a meteorite isn't very practical.

We know that meteorites can fall anywhere at any time. They are as likely to crash into your mailbox (this has happened to people) as anything else. Meteorites that are seen falling are called falls. Those that are not observed but are eventually found are called—you guessed it—finds. Finds represent two-thirds of all known meteorites. You don't have to see a meteorite fall to hope to find one. But it helps to know where to look. You wouldn't head to downtown Cleveland to see elephants in the wild or to Florida to do some snowboarding, so it makes sense to ask: Where have meteorites been found already? Have all the meteorites been found at existing craters?

You can't hunt for meteorites at Meteor Crater today because of legal restrictions. But there are lots of places you can look, and it makes good sense to go where they've been found before. Many meteorites have been discovered in old fall areas. So you can do some homework to identify likely spots. This is much easier than it used to be. There are websites (see the Resources section) that list meteorites by state and country. When you look at those websites, you'll probably be surprised at how many meteorites have been found in your area—or how few. There are only 6 recognized meteorites in Oregon. There are 108 in Arizona. Hmm. Why do you suppose that is?

Climate and land cover are the key. The unfortunate reality is that a humid, marshy, or forested region with thick underbrush, such as western Oregon, is not a good place to hunt for meteorites. Dense vegetation hides the ground. You could be walking over meteorites and never know it. Likewise, places where the ground is very soft aren't good because a stone falling from space can penetrate the soil on impact and be covered up. Also, places where the rocks are dark, such as on the basalts of eastern Oregon, are bad places to look for dark rocks.

The regions where most meteorites are found are deserts, both hot and cold, and farmland where the soil is frequently plowed. Many meteorites have been discovered in rock piles along the edges of cleared fields throughout the midwestern United States. The desert areas of the southwestern United States, including parts of California and Arizona, are excellent hunting grounds. Vegetation is sparse, the climate is dry, and the surface of the ground may have remained basically unchanged for thousands of years. Dry lakes are also good places to look. Wherever you look, you have to become familiar with local rocks first so you can look for something different. This takes practice.

Flat, featureless terrain is the best place to search. It's helpful if the color of the surface is light so the dark fusion crust on a meteorite will stand out against the background. It's even more helpful if there is a mechanism that brings buried rocks to the surface. Such a place exists: Antarctica. Meteorites have been found there since exploration of the continent began. The first was found in 1912. Since then more than twenty thousand specimens have been recovered.

Antarctica is covered by an ice sheet thousands of feet thick that is white, flat, and mostly devoid of native rock. Meteorites falling there are well preserved because it is cold and dry and they are quickly covered by snow or ice and protected. Glaciers move outward from the middle of the continent toward the coasts. When glacial ice encounters mountain ranges, it piles up along the slopes where fierce, dry downhill winds cause the ice to evaporate, freeing any buried meteorites. Movement of the ice along the mountains tends to concentrate the meteorites into areas where many have been found together. Most are located in very old blue ice, which is kept free of snow by the constant winds. Measurements of radioactive isotopes in the meteorites show that some have been resting there for millions of years.

Before you pack your bags, be aware that Antarctica is not only bitterly cold, it's also practically inaccessible and totally off limits for personal meteorite hunting. The Antarctic Treaty, which became effective in 1961, includes a recommendation that protects all geological specimens, including meteorites, from private collection. However, the Antarctic Search for Meteorites program, administered by the National Science Foundation, is allowed to collect Antarctic meteorites for scientific study.

Okay then, how about someplace warmer? Since the year 2000, there has been a tremendous surge in the discovery of meteorites in the hot deserts of the world, and this is good news. More than five thousand new meteorites have been found in Northwest Africa alone. Many are picked up by nomads and sold to museums and collectors. Some important finds have occurred in this way, including specimens from Mars and the Moon. The many new discoveries have also made ordinary meteorites less expensive to purchase.

IS METEORITE HUNTING LEGAL?

In the United States, meteorites belong to the landowner. If you want to search on private property, you must first get permission to be on the land and look for them. You must also ask for permission to keep what you find. A lot of land belongs to the government in one form or another, such as state parks, national forests, and so on. Before searching on public land, check regulations that apply there. Meteorite hunting is forbidden on some public lands, including national parks. On other public lands, it may be okay. And even if you get permission or it is legal to hunt for meteorites, it's important to respect the land. Never leave any trash, open holes where you removed rocks, or any other signs that you were there. The actions of a few careless people have caused entire meteorite collecting areas to be closed.

In the United States it is completely legal to buy, own, and sell meteorites. In some other countries people aren't so fortunate. Canada and Australia have strict rules about exporting meteorites, and India allows no private ownership of meteorites at all. In some countries, meteorite hunters have been jailed for collecting. Not surprisingly, countries that restrict collecting have very few recovered specimens.

We haven't mentioned what meteorites are worth because it's hard to say. If they are from Mars or the Moon, they are extremely valuable. But ordinary chondrites, especially those from Northwest Africa, can be inexpensive. Meteorites from observed falls like Park Forest are expensive because they are historic.

WHAT TO TAKE ON YOUR METEORITE HUNT

If you venture into unpopulated areas, take water, food, a first aid kit, adequate clothing, and a charged cell phone. However, don't count on the cell phone in case of an emergency; you may not have reception. Watch out for snakes. Don't go alone, and do tell people where you're going. A sudden shower can turn a dry wash into a river and a dry lake into mud in minutes. Be sure to keep your wits about you while you're concentrating on finding those elusive black rocks.

To distinguish promising meteorite candidates from ordinary black rocks, you'll need a small 10x magnifier and a good magnet. Don't just grab something off the refrigerator door. It needs to be much stronger. Even stony meteorites can contain

There is a meteorite in this picture. Can you guess which rock it is? It has a shiny black crust and a blocky shape, a very typical appearance.

enough metal to be attracted to a magnet, if only weakly. Some people attach a magnet to a cane or stick to make it easier to pick up promising rocks.

METEORITES ARE WHERE YOU FIND THEM

Let's review what you need to look for to find a meteorite. You're looking for rocks that are generally heavier than normal rocks of the same size because they contain some iron. You're looking for a very dark fusion crust. If the rock is broken, you may see a lighter interior. Metal grains and tiny spherical chondrules will really make your day. Use your magnifier to examine potential meteorites carefully. They will probably be attracted to your magnet. You may find a chunk of iron, which will definitely stick to a magnet, but it may not be a meteorite. Further testing is necessary to determine whether nickel is present. And even if the rock contains nickel, more laboratory tests are required to prove it's a meteorite.

Weather is the greatest enemy in identifying meteorites in the field. A meteorite that has been exposed to the elements for a long time may be unrecognizable. If you use a metal detector, you can still hope to locate one of these. It depends on how much metal is in it, how weathered it is, and if it's buried.

The best way to find a rock that's different is to know as much as you can about the area. Just because a rock seems like it wasn't there yesterday doesn't mean it dropped in from space overnight. If you want to be sure a rock is a meteorite, you'll eventually have to submit it to a laboratory for analysis. If you want it to be recognized as an official meteorite, you'll have to have it inspected by someone knowledgeable and then it must be authenticated by a lab before being submitted to the Nomenclature Committee of the Meteoritical Society for official recognition. This involves cutting and giving up part of the meteorite. Appendix C can help you get started.

But first you need to do everything you can to eliminate other possibilities. A good way to learn to recognize real meteorites is to buy a few inexpensive ones and examine them carefully.

Through the years we have seen an amazing number of false meteorites, often called "meteorwrongs." Some of them were really convincing. If you believe you've found a genuine meteorite, compare it with photos of common "meteorwrongs."

Sometimes a meteorite stands out against the surrounding rocks, like this one in Nevada. —Photo courtesy Sonny Clary

Sometimes you really have to dig for treasure, and you won't find it without a metal detector. Geoff Notkin worked hard for this Brenham pallasite from a famous meteorite location in Kansas. —Photograph by Geoff Notkin © Aerolite Meteorites, www.aerolite.org

Author O. Richard Norton hunting meteorites with a metal detector in Arizona. No luck that day.

Persistence does pay off. Matthew Maidment was disappointed when he didn't find a meteorite in the desert right away, but after searching for a while in an area where they have been found before, he yelled, "I got one! I got one!" It was his first meteorite discovery and a very nice stone that sticks to a magnet. —Photo courtesy David Maidment

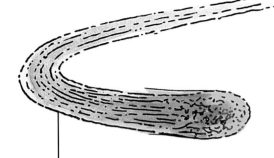

METEORITE HUNTER'S CHECKLIST

If it's a stone:

1. Does it have a black or dark brown fusion crust? (Don't confuse this with a surface stain.) The crust should be less than 1 millimeter thick.

2. Does it have flow lines?

3. Is it heavier than a normal rock of its size?

4. Is it attracted to a magnet?

5. Does it have "thumbprint" pits? Some stones do.

6. If it's broken, is the interior lighter? If it's unbroken, you may want to file off the surface in a small area so you can peek inside.

7. Can you see small spheres (chondrules) or metal grains?

If it's an iron:

1. Is it attracted to a magnet?

2. Is it torn and distorted, with sharp edges? Is it lumpy?

3. Does it have "thumbprint" pits?

4. Does it have flow lines?

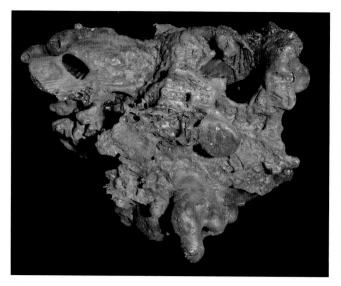

This rock, made of fused soil from a very hot fire in a forest, was completely melted. Meteorites are never completely melted.

Slag, a general term for waste products from the process of separating metal from ore, comes in many forms, and some can fool you. This piece even seems to have flow lines and "thumbprints," but it's not attracted to a magnet.

This sedimentary rock is full of spheres, but they are not chondrules. At 5 millimeters in diameter, these spheres are too large to be chondrules, and there isn't material filling the spaces between them.

Botryoidal hematite, an iron oxide mineral, is sometimes mistaken for a meteorite. The spheres aren't separate like chondrules and look more like soap bubbles or a bunch of grapes.

These spheres aren't chondrules either. They're from an outcrop of a type of volcanic rock known as spherulitic rhyolite, in a road cut in Yellowstone National Park.

It may look like this weathered desert rock has a fusion crust, but the black coating is too thick—more than 3 millimeters. It is probably a coating of iron oxide.

Sometimes nature likes to play tricks on us. Hematite concretions, like this one from Utah, formed when iron oxide minerals precipitated out of groundwater long ago. Concretions are often mistaken for eggs, turtle shells, bones, fossils, and meteorites.

METEORITES ROCK!

1

2

3

4

5

If your dad is a professional meteorite hunter, well . . . you do have an advantage.
1. Nick Garcia, Hopper the meteorite-hunting dog, and dad Ruben Garcia with meteorites and metal detector.
2. Ruben Jr. and Nick follow their dad on the trail at Holbrook, Arizona.
3. Ben Garcia on a meteorite-hunting trip. Check out the license plate.
4. Nick Garcia finds a Nevada meteorite.
5. Ruben Jr. picks up hundreds of meteorites in California.

—Photos courtesy Ruben Garcia, MrMeteorite.com

STARTING A COLLECTION

If you can't wait to find your own meteorite, you can always buy one. Meteorites are often sold at gem and mineral shows and at rock shops. Check online auction sites and dealers' websites. As always when collecting, you must do your homework. There are some honest people who think they're selling meteorites, but they actually have "meteorwrongs." Others are not so honest. Look for the insignia of the International Meteorite Collectors Association (IMCA), and visit the association's website for great information on all things meteoritic. They have a lot of good advice on making a meaningful collection.

Look for these logos when buying meteorites online. An IMCA member will have an official number in the box.

And why collect meteorites? Don't do it just because they're cool (they are), they come from space (they do), and sometimes they're worth a lot of money (also true). Collect meteorites so you can have a chance to hold one in your hand and think about where it came from and how long and how far it traveled to get here. Think about the stories it has to tell: stories about the early years of the solar system, violent and cataclysmic impacts, and the bitter cold of space. Think about the heavy elements born in the death throes of giant stars and how all of that led to us, here, in this beautiful world filled with warmth and life. Think about the mysteries still hidden within these ancient rocks.

TOP TEN IMPACT CRATERS OF THE WORLD

Name	Location	Diameter (miles)	Age (millions of years)
Vredefort	Free State, South Africa	186	2,020
Sudbury	Ontario, Canada	155	1,850
Chicxulub	Yucatán, Mexico	106	65
Manicouagan	Quebec, Canada	62	214
Popigai	Siberia, Russia	62	35.7
Chesapeake Bay	Virginia, United States	56	35.5
Acraman	South Australia, Australia	56	590
Puchezh-Katunki	Nizhny Novgorod Oblast, Russia	50	167
Morokweng	Kalahari Desert, Northwest Province, South Africa	43	145
Kara	Nenetsia, Russia	40	70

TOP TEN IRON METEORITES

Meteorite	Place and year of discovery	Weight (metric tons)
Hoba	Otjozondjupa Region, Namibia, 1920	60
Campo del Cielo (El Chaco)	Chaco, Argentina, 1969	37
Cape York (Ahnighito)	Greenland, 1894	31
Armanty	Xinjiang, China, 1898	28
Bacubirito	Sinaloa, Mexico, 1863	22
Cape York (Agpalilik)	Greenland, 1963	20
Mbosi	Mbeya Region, Tanzania, 1930	16
Campo del Cielo (La Sopresa)	Chaco, Argentina, 2005	15
Willamette	Oregon, USA, 1902	15
Chupaderos I	Chihuahua, Mexico, 1852	14

But wait, what about stones? They usually break up before they reach the ground. The largest recorded stony meteorite (an H5 chondrite) fell in Jilin, China, on March 8, 1976. It weighed about 4 tons. A 1-ton achondrite stone fell in Norton County, Kansas, in 1948.

IF YOU FIND A METEORITE

If you think you may have found a meteorite, the first step is to rule out the possibility that it's an Earth rock. Here are some resources that can help.

Washington University in St. Louis, Department of Earth and Planetary Sciences:
meteorites.wustl.edu/meteorwrongs/meteorwrongs.htm
An excellent guide, with photos of numerous "meteorwrongs" and explanations of why the objects aren't meteorites.

International Meteorite Collectors Association: www.imca.cc
Look under "Met Info" and then under "Finding Meteorites" for a quiz with photos and explanations.

Meteorite-identification.com: meteorite-identification.com
Information on simple tests that can help determine whether an object is a "meteorwrong," along with links to many good sources about identification (meteorite-identification.com/idweb.html) and how to get your rock tested (meteorite-identification.com/verification.html).

If you believe you may have a meteorite and want to get an official name and classification, you must submit a sample for analysis. A local planetarium, reputable meteorite dealer, or collector can provide an initial examination. Be sure to make an appointment first. If your rock is thought to be a meteorite, contact the nearest meteorite laboratory (usually associated with a university planetary science, geology, or astronomy department) and ask permission to send the stone. Never send a sample without making prior arrangements. It may not be delivered to the right person or you may have the wrong address. Be sure to include return postage if you want the specimen sent back to you. Also include as much information as possible about when and where the stone was found. Documentation is very important.

RESOURCES

Books

Bevan, Alex, and John de Laeter. 2002. *Meteorites: A Journey through Space and Time.* Washington, DC, and London: Smithsonian Institution Press.

Carion, Alain. 2009. *Meteorites.* Paris, France: Alain Carion.

Norton, O. Richard. 1998. *Rocks from Space.* Missoula, MT: Mountain Press Publishing Company.

Notkin, Geoffrey. 2011. *Meteorite Hunting.* Tucson, AZ: Aerolite Meteorites.

Reynolds, Mike D. 2010. *Falling Stars: A Guide to Meteors & Meteorites*, second edition. Mechanicsburg, PA: Stackpole Books.

Smith, Caroline, Sara Russell, and Gretchen Benedix. 2009. *Meteorites.* Buffalo, NY: Firefly Books.

Spangenburg, Ray, and Kit Moser. 2002. *Meteors, Meteorites, and Meteoroids.* New York: Franklin Watts.

Magazines

Meteorite: www.meteoritemag.org

Meteorite Times: www.meteorite-times.com

Online Resources

American Meteor Society: www.amsmeteors.org

Information on meteors, meteor showers, and more, and an online form for reporting bright fireballs (www.amsmeteors.org/fireball2/form2.php).

Antarctic Search for Meteorites Program: geology.cwru.edu/~ansmet/index.html

Information on the search for meteorites in Antarctica, with photos and field reports.

International Meteorite Collectors Association: www.imca.cc

Information about meteorites and guidance in meteorite collecting.

The Meteorite Exchange Network: www.meteorite.com

Listings of meteorites for sale and meteorite dealers, information on meteorites, photographs, and meteorite news.

The Meteoritical Society: meteoriticalsociety.org

Many resources on meteoritics and planetary science, including a database of all known meteorites, with many search options (www.lpi.usra.edu/meteor).

NASA Solar System Exploration website: solarsystem.nasa.gov/index.cfm

Scientific and educational information on the solar system and NASA missions, including multimedia presentations. For information on meteorites, click on "Planets." Also includes a section just for kids (solarsystem.nasa.gov/kids/index.cfm)

GLOSSARY

ablation. The removal of material by heating and vaporization as a meteorite passes through Earth's atmosphere.

accretion. The gradual accumulation of material through collision of particles in the solar nebula.

achondrite. The class of stony meteorites lacking chondrules.

asteroid. A large body of rock, metal, or rock and metal, smaller than a planet but larger than a meteoroid, orbiting the Sun, usually within the asteroid belt.

asteroid belt. A region between the orbits of Mars and Jupiter where most asteroids are located.

basalt. An igneous rock formed by erupting volcanoes on the surface of Earth, the Moon, Mars, Venus, and the asteroid Vesta.

breccia. A rock consisting of angular fragments cemented together.

carbonaceous chondrite. The most primitive class of stony meteorites; many contain organic compounds.

chondrite. The class of stony meteorites containing chondrules.

chondrule. Spherical bodies usually measuring 1 millimeter or less in diameter found in chondrites. Chondrules form through melting and recrystallization of minerals in the solar nebula.

comet. A small body composed of rock and ice in orbit around the Sun. When it nears the Sun it forms a cloud of gas and dust called a coma and a long tail.

differentiation. The process by which a planetary body melts and separates into layers of different density and composition, usually forming a core, mantle, and crust.

element. A substance that cannot be reduced by normal means to anything else. For example: hydrogen, helium, and iron.

fireball. A very bright meteor.

fusion crust. A dark, glassy coating that forms on stony meteorites as they are heated in the atmosphere.

igneous rock. Rock formed from the cooling of magma, either on the surface or deep underground.

inner planets. The four smaller planets closest to the Sun—Mercury, Venus, Earth, and Mars—which are primarily made of rock and metal.

isotope. Atoms of the same element with different atomic masses, some of which are unstable.

meteor. The light produced when a meteoroid enters Earth's atmosphere and burns due to friction.

meteorite. A meteoroid of rock, metal, or rock and metal that has fallen on the surface of a planet or moon. A **meteorwrong** is a rock from Earth that looks a bit like a meteorite and is mistaken or misrepresented as a meteorite.

meteoriticist. A scientist trained in chemistry and mineralogy who studies meteorites.

meteoritics. The science of meteorites and the origin of the solar system.

meteoroid. A chunk of rock, metal, or rock and metal, smaller than an asteroid, orbiting the Sun.

meteor shower. An annual event in which many meteors seem to originate from a single point in the sky, caused by Earth's passage through a cloud of debris, usually cometary.

nebula. A cloud of dust and gas in space.

orbit. The path of one object around another in space.

ordinary chondrites. The most common type of stony meteorites that contain chondrules.

outer planets. The four giant planets beyond the asteroid belt which are made of gas and ice.

parent body. A planet or asteroid that has pieces broken from it during collision, producing meteorites if the pieces travel to Earth.

radioactive. The unstable form of an element that gives off radiation when it decays.

sedimentary rock. Rocks formed from weathered pieces of preexisting rocks.

spectrum (plural: spectra). The range of variation of something that can change from one end to the other—for example, the colors of the rainbow.

strewn field. The generally elliptical area containing meteorites from a meteorite shower, where many stones fall at the same time.

tektite. A glassy body formed in a major impact when melted rock from Earth's surface is ejected into space and then reenters the atmosphere, often having an aerodynamic shape.

thin section. A slice of rock 0.03 millimeters thick on a glass slide used to study its minerals with a microscope.

vaporize. To change a solid or liquid into a gas.

Widmanstätten pattern. The pattern of crystal plates of mineral alloys in octahedrite iron meteorites, revealed by etching.

GENERAL INDEX

Page numbers in bold refer to photographs.

shock wave, 23, 56, 63, 64, 66
Shoemaker, Eugene, 66
shrapnel, **49**, 59, 63, 64
Sikhote-Alin, **56**, **58**, **59**, 62, 64
silicate minerals, 19
Simon, Steven, 6
slag, 78
solar system, 11–13, 18; formation of, 23–25; small bodies in, 18
spectra, 18; reflectance of, 19, 39
Spirit, **45**
Steins, **21**
stony-iron meteorites, 28, 29, 50, 51
stony meteorites, 20, 23, 28, 30–33, 59, **74**
strewn field, 60
s-type asteroids, 19–20
Sun, 1, 11–13, 15, 18, 23–25, 34, 53
supernova, 23, 24
Sylacauga, 67

taenite, 48
tektites, **68**
telescopes, 14

thin section, iv, 27
Thomson, William, 48
thumbprints, **58**, 59, 63
Toutatis, **21**
Trojans, 16

University of Western Ontario, 8, 9

Vaca Muerta, **50**
Valera, 67
Venus, 11, 17, 25
Vesta, 17, **19**, 26, 29, 38, **39**

water, 36
Widmanstätten, Count Alois Beckh von, 48
Widmanstätten pattern, 48, **49**, **51**
Winslow, Arizona, 64

Yeomans, Donald, 53
Yucatán Peninsula, 69

Zagami, **43**

METEORITE INDEX

ASTEROID INDEX

—Photo courtesy Phillis Temple

O. Richard Norton fell in love with meteorites when he studied astronomy at UCLA with renowned meteoriticist Frederick C. Leonard. As director of the Fleischmann Planetarium at the University of Nevada at Reno and the Flandrau Planetarium at the University of Arizona in Tucson, he taught astronomy and shared his enthusiasm for meteorites, geology, and photography in public lectures and community education classes. He traveled to Cape Canaveral to film the *Apollo* launches, designed a fish-eye motion picture system that flew on the space shuttle *Challenger*, and led field trips to photograph total solar eclipses and comets around the world. His previous books about meteorites are *Rocks From Space*, *The Cambridge Encyclopedia of Meteorites*, and *Field Guide to Meteors and Meteorites*. He was a fellow of the Meteoritical Society and a contributing editor of *Meteorite* magazine.

Dorothy Sigler Norton is an artist and scientific illustrator. She studied art at Washington University in St. Louis and the University of Iowa. Her colorful illustrations and ink drawings have appeared in many magazines and books, and her large paintings hang in the national geological museums in Japan. With Richard, she operated Science Graphics, a company that supplied science teaching materials to universities worldwide. Her passion for meteorites began when she discovered it's actually possible to own one, and she's been collecting and searching for them ever since. She is a member of the International Meteorite Collectors Association and serves on the editorial advisory board of *Meteorite* magazine.

Richard and Dorothy have asteroids named for them: 163800 Richardnorton and 149243 Dorothynorton.